013909225

AF616102

WITHDRAWN
FROM STOCK
For conditior

Vomeronasal Chemoreception in Vertebrates

A STUDY OF THE SECOND NOSE

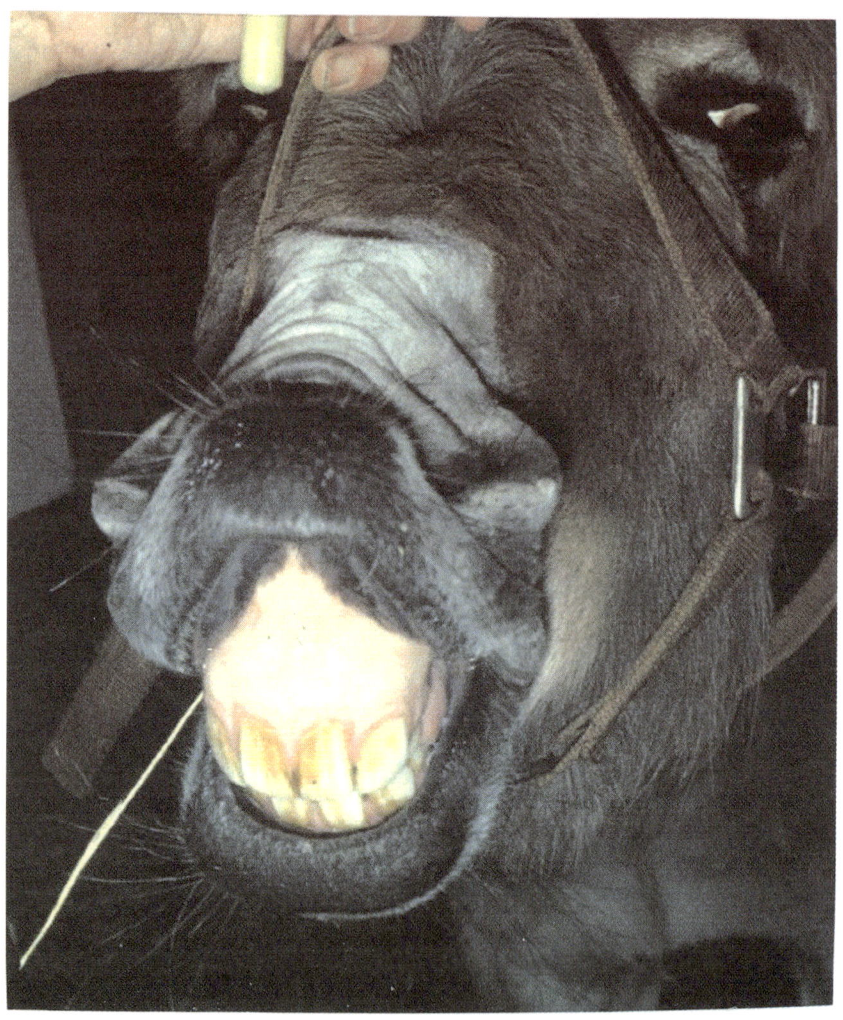

Flehmen in stallion: response to estrous urine — elicited by sniffing of volatiles and or nonvolatiles (courtesy Fay Lindsay©)

Vomeronasal Chemoreception in Vertebrates

A STUDY OF THE SECOND NOSE

Charles Evans
Glasgow Caledonian University, Scotland

Foreword by:

D. Michael Stoddart
Chief Scientist
Australian National Antarctic Research Expeditions

ICP
Imperial College Press

Published by

Imperial College Press
57 Shelton Street
Covent Garden
London WC2H 9HE

Distributed by

World Scientific Publishing Co. Pte. Ltd.
5 Toh Tuck Link, Singapore 596224
USA office: Suite 202, 1060 Main Street, River Edge, NJ 07661
UK office: 57 Shelton Street, Covent Garden, London WC2H 9HE

Library of Congress Cataloging-in-Publication Data
Evans, Charles.
Vomeronasal chemoreception in vertebrates : a study of the second nose / Charles Evans.
p. cm.
Includes bibliographical references and index.
ISBN 1-86094-269-5 (alk. paper)
1. Chemical senses. 2. Jacobson's organ. 3. Pheromones. 4. Pheromones--Receptors. I. Title.

QP455 .E936 2003
573.8'7716--dc21

2002038701

British Library Cataloguing-in-Publication Data
A catalogue record for this book is available from the British Library.

Printed in Singapore.

Dedication

to

Robert W. Goy
(1924–1999)

y meistr

ACKNOWLEDGEMENTS

With thanks for illustrations reprinted with permission from the Institutions and Companies given below.

Cover: {front} Library, Trinity College, Dublin, 'The Book of Kells': TCD Mss 58; fol 76v (1. 13-17, end detail, right) entitled "*Blue wolf, tongue extended*".

Amer Assoc Adv Science: 2.1,b; 5.13; **Aust Mammal Socy**: 2.6,a/b; 7.6,a/b; **Brill Acad Publ** 7:2 a/b; 7.4,a; 7.6,c; 7.8; **British Museum Natural History**: 1.3; **Cambridge Univ Pr**: 2.12,a/b; 2.13,a; 4.1; 4.2,c/d; **Cell Pr Inc** 6.5,b; **Chapman & Hall**: 1.4; **CRC Pr Inc**: 3.3; 2.18,a; **Elsevier Science**: 1.2; 2.13,b; 3.2; 4.1; 4.3; 5.4; 5.16,a; 6.5,a; 7.1; 7.6,e; 7.10,a/b. **G Fischer**: 2.2,a; **Glasgow Univ Library**: 2.9,a; 2.10,a; **Glasgow Caledonian Univ Library** 7.9,a; **Greenpeace**: 7.6,g; **Kluwer/Plenum Pr**: 1.2; 2.1,a; 2.19,b/c; 3.2; 4.4,a; P 3.1,A/B; P.4,A; 5(Hd.); 5.4; 5.14,a; 7.6,d; **Masson et Cie**: 7.2,b; **Museum Natl d'Hist Naturel, Paris**: 2.11,a; 5.8,b; 7.4,b; **Natl Acad of Sciences,Wash DC**: 6.5,a; **Nature**: 6(Hd.); 5.5,a; 5.14,b; **New York Acad Sciences**: T.3.1; 4.4,a; 5.9,b; 6.2; 6.4; **Oxford Univ Pr**: P.2,C/2; P.5,1; P.6,A; 2.16,a; 3,b; 5.3,b; 6.5,a; 7.2,c; **Pergamon Pr**: 2.19,a; **Rockefeller Univ Pr**: 5.8,a; **Soc d'Endocrinol**: 4.4,b; **Soc. Study of Reproduction**: 7.10,c; **Smithsonian Inst**: 1(Hd.); 3(Hd.); 7.9,b; **Springer Verlag**: 2.4,b/c; 4.5; **Univ Stellenbosch**: 2(Hd.); **Urban & Fischer Verlag**: 2.4,a/c; **John Wiley/Liss Co**: 2.4,b; 2.7; 2.10; 2.11,b; 4.6,a; P.7.1;

[Hd. = heading; P = plate; T = table].

And for permission to use quotations from their publications:-

Ch.1, Allen & Unwin: 'The Biochemistry of Genetics' (Haldane, 1954); **Ch.4**, Penguin Books: 'A Case of Knives' (McWilliam, 1988);

Ch.6, J. Calder: 'The Motor Show' Plays, vol. V. (Ionescu, 1960); other sources are:- **Ch.7**: Darwin C. (1839) 'The Voyage of the Beagle', J. Murray Lond., and: Oxford English Dictionary (compact edit. vol.I, 1979).

My grateful thanks also to all those who variously helped with laboratory and other investigations, as well as with critical improvements to the presentation.

CONTENTS

FOREWORD

Thanks to the keen eyes of a Danish military surgeon, and the burning of some late-night oil, the world of comparative anatomy came to know of the existence of the vomeronasal organ (VNO), sometimes known as Jacobson's organ, after its discoverer Ludwig Levin Jacobson. Despite almost two centuries of advancement of anatomical knowledge, rather little is known about the organ, even to this day. The eponymously-named organ was thought by Jacobson to be secretory in function but this is now known not to be the case. Its location, in the nasal cavity at the base of the nasal septum where that piece of cartilage joins the vomer bone, suggests it has something to do with the nose and the sense of smell. And so it has. The organ certainly does process molecules plucked from the outside world but the transmission of the information gleaned enters the brain not *via* the olfactory bulb, but *via* a small structure closely appressed to the olfactory bulb, and known as the accessory olfactory bulb. Vomeronasal nerves thus run directly to the evolutionarily ancient smell-brain, or rhinencephalon, *via* their own pathways. If the sense of smell is, by definition, mediated by the nose and nasal mucosa capable of sensing compounds of high volatility, then the VNO is not part of the sense of smell. But it is part of the wider chemical sensory systems found in the air-breathing vertebrates.

The sense of smell, with which we can enjoy a bursting rosebud on a summer's day, or a fine Cabernet, is a relatively recent evolutionary acquisition. For most of evolutionary time our ancestors did not breathe air and gained information about their surroundings by "tasting" their aquatic environments. The so-called "olfactory rosettes" in the heads of fishes are well-known to anyone who has cleaned a fish; they exemplify the sort of environmental chemical sensory apparatus that supported animal evolution for many millions of years.

When animals fully rid themselves of the necessity to live in water, and started to breathe air, the VNO flourished along with the nose and true olfactory system. The organ is best-developed in reptiles and mammals and in the latter class is almost universally present. Perhaps the main reason for our rather poor understanding of the organ is because our own species, *Homo sapiens*, is amongst the few that appear to be lacking a VNO. There is some debate about this point, though there is agreement that a VNO starts to develop in the human embryo. While some anatomical traces of the organ can be found in some adults, there is as yet no unequivocal proof that a fully functional organ actually exists in humans.

In this book Charlie Evans seeks to demystify the VNO by bringing together a considerable amount of information about the organ's evolution, functional morphology, development, activation and neuroendocrinology, and the genetic requirements of accessory chemoreception. He deals also with the nature of chemical signals that convey information about one individual to another, and about the behaviours animals express when pumping their VNOs with chemically laden aerosols. It clearly sets what is known about the organ and, perhaps more importantly, where the gaps are in our knowledge. The VNO is attuned to molecules of low volatility that cannot trip the nose, enabling the organ to respond to large molecule chemical compounds that advertise (eg.) the state of sexual readiness of the advertiser. Considerable experimental work with rodents has shown that the organ is able to detect the sexual status of a potential partner, though this does not obviate the role of the principal olfactory system in sexual behaviour. The VNO's function can be regarded as "priming" the individual for reproduction. Much more is yet to be learned about how the organ functions, and in which biological contexts it exerts its influence.

It is to be hoped that this monograph will encourage a new generation of scientists to take up the challenge and add their contributions to our understanding of this small but important anatomical structure. Chirurgien-major Jacobson would have been delighted that the tiny seed he planted has grown into a substantial tree.

Michael Stoddart
Hobart, November 2002

ABBREVIATIONS

AOB	Accessory Olfactory Bulb
AOB-x	Accessory Bulbectomy
AOS	Accessory Olfactory System
APH	Aphrodisin
BNST	Bed Nucleus Stria Termialis
BT	ButyldihydroThiazole
CNS	Central Nervous System
DB	Dehydo-*exo*Bevicomin
EOG	ElectroOlfactogram
EVG	ElectroVomeronasogram
F.	Flehmen
GPCR	G-Protein Coupled Receptor
GnRH/LHRH	Gonadotrophin Releasing Hormone/Luteinising Releasing Hormone
JO	Jacobson's Organ
LOT	Lateral Olfactory Tract
MHC	Major Histocompatibility Complex
MOB-x	Bulbectomy
MOE	Main Olfactory Neuroepithelium
MOS	Main Olfactory System
NO	Nitric oxide
N-Pp/N-Pd	Naso-Palatine Papilla/Duct
NT	Terminal Nerve (*Nervus terminalis*)
OBP	Odourant Binding Protein
OR	Olfactory Receptor (molecule)
ORN	Olfactory Receptor (neurone)
OCAM	Olfactory Cell Adhesion Molecule
SO/MO	Septal Organ (of Masera)
VNd	Vomeronasal Duct
VNE	Vomeronasal Neuroepithelium
VMod	Vomeromodulin
VNO	Vomeronasal Organ
VNOR	Vomeronasal Olfactory Receptor (molecule)
VNR/r	Vomeronasal gene/allele

VRN	Vomeronasal Receptor (Neurone)
VNS	Vomeronasal System
VN-x	Vomeronasalectomy
T	Testosterone
UHF	Ultra-High Frequency

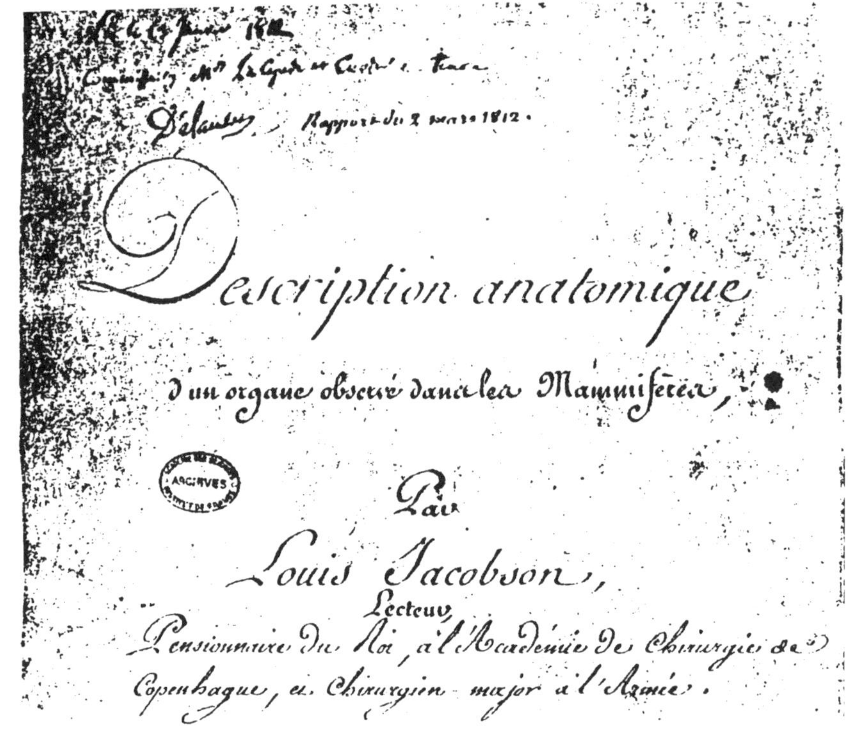

Rapport du 2 mars 1812.

Description anatomique

d'un organe observé dans les Mammifères,

Par

Louis Jacobson,

Lecteur,

Pensionnaire du Roi, à l'Académie de Chirurgie de

Copenhague, et Chirurgien-major à l'Armée.

Fig. Int.1 Title page of Mémoire by L.L. (Louis) Jacobson (submitted to l'Institut de France in 1812).

INTRODUCTION

The discovery of the structure we now call the vomeronasal organ emerged from comparative anatomical work very much of its time. It was named after the veterinarian and physician Ludwig Levin Jacobson (1783–1843) who first described the gross morphology of the structure in the noses of some 15 genera of mammals, and provided figures showing details of the nerve supply to the organ. The work emphasised the "new" organ's secretory capabilities, and commented on its connections with those between the mouth and nose. Jacobson initially presented his findings in a monograph to the veterinary audience (Jacobson, 1813). The publication history of his original work (in Danish) is documented by Trotier and Døving (1998). This gives a translation, together with reproductions of his striking drawings of the horse nose. The other domestic and non-domestic mammals he examined and/or dissected were also discussed in a "mémoire" apparently sent to the Académie de France. The facsimile page (Fig. Int.1, opposite) of Jacobson's submission (dated 12 January, 1812) bears the names of its sponsors: de Lacepede, Tenon and Georges Cuvier under whose name the "Rapport fait à l'Institut, sur un Mémoire de M. Jacobson" appeared (Cuvier, 1812; transl. in Bhatnagar and Reid, 1996). An illustrated version of this text was later re-issued in a sparsely distributed paper, edited by Hollnagel-Jensen and Andreasen (1948). The original manuscript of the mémoire is thought to have been lost (?) from the Parisian archives, but does reside in a Copenhagen library (Trotier and Døving, 1998). Although mentioned, but not considered by Jacobson, the first record of such a nasal structure appeared in a much earlier treatise by Frederick Ruysch (1703). While this shows a nasal septum [Fig. Int.2(a)] in which a probe (D) undoubtedly demonstrates the existence of an aperture and a duct-like "nasal canal" in the human neonate, it does not justify any priority in the discovery of the organ. The figure legends in Ruysch's commentary do no more than label it (E) as a part of septal anatomy; he makes no functional attribution, nor does he recognise it as a vertebrate characteristic. One other precedent to Jacobson is in a treatise on the "olfactory organ" by Sœmmerring (1809, not seen).

The nomenclature of the organ is determined by anatomical convention, regrettably discarding the practice of identification after its describer. Its official title is "*Organ on Vomeronasale Jacobsonii*" but

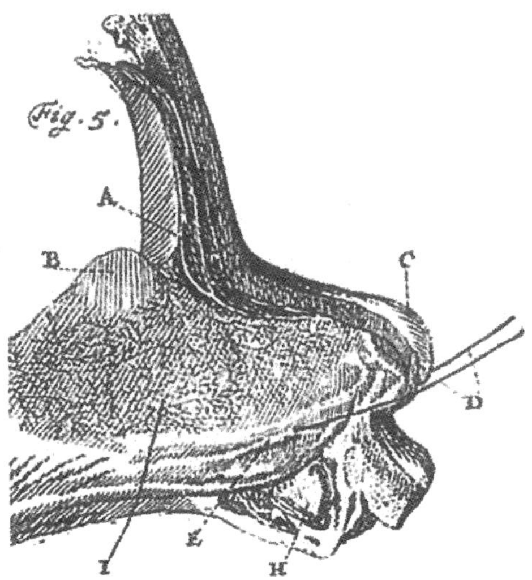

Fig. Int.2(a) "*Neonatus*" Fig. v from: Fred. Ruysch (1703) *Thesaurus Anatomicus*: *fasc.* III, Tab. 4; [p. 49]. Jans Wæsberg, Amsterdam.

the working name is the "Vomeronasal Organ", occasionally qualified as "Vomeronasal Organ (of Jacobson)". The persistence of the original eponymous label is largely due to the quirk of continued usage amongst researchers on snakes and lizards (e.g. Young, 1993); although mammalian specialists too will occasionally relapse (Jastrow and Oelschlager, 1998). The name itself derives from the apparent plough-like shape of a medio-basal nasal bone: "By bony apposition caudally the U-shaped vomer gradually changes into a Y-shape" (Sandikcioglu *et al.*, 1994).

Following Jacobson, the morphological and histological studies in the 19th century did establish that the distribution of the organ was from amphibia upwards, and that in general its microscopic structure was relatively invariant (Herzfeldt, 1888; von Mihalkovics, 1898). Its presence in humans was confirmed for pre-natal stages, and its occurrence again demonstrated, this time in adults, by the insertion of probes (sounds) into the septum [Fig. Int.2(b)] of cadavers and patients (Gratiolet, 1845; Potiquet, 1891). Much of the subsequent speculation

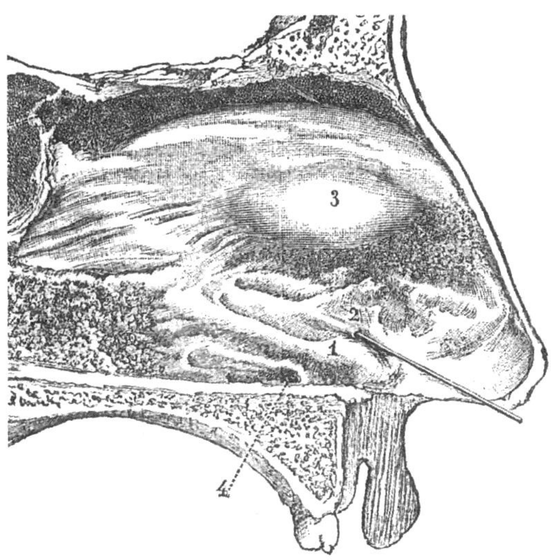

Fig. Int.2(b) "*Cloison des fosses nasales*" Fig. 1 from: Potiquet (1891). Du canal de Jacobson. de la possibilité de le reconnaître sur le vivant…etc… *Rev laryngol. d'otol. rhinol.*, Paris, **24**, 737–753.

on function did not go beyond Jacobson's conclusions that it could be a sensory organ, possibly with some secretory capabilities. Indeed, most attention was given to the evolutionary and taxonomic importance of the nature of and variants in the cartilage surrounding the organ.

The recent resurgence of interest in smells and scents has considerably expanded information on, if not a full understanding of, the multiple chemoreceptive systems in the nose. New techniques and more systematic approaches have increasingly been applied to the erstwhile Cinderella area of the chemical senses.

A comparison of two compendia on methodology, Moulton *et al.* (1978) with Spielman and Brand (1995), illustrates the refinements of investigative procedures.

The subject of this book is one of several minor chemoreceptive systems which together merited only limited consideration in previous studies on general olfaction by Halaz (1990) and by Farbman (1992). Over the last decade, the vomeronasal sub-system has emerged as a distinct area of interest within the biology of chemoreception. Although specialist reviews regularly appear on vomerolfaction, a particular study of this system is warranted by the increasing recognition given to accessory olfaction.

Comparative approaches to accessory olfaction have been consistent amongst reptilian specialists (Wright, 1883; Kahmann, 1932; Liu, 1999). Until the 1960s, mammals were not quite so fortunate in their devotees. A notable early exception was another Glaswegian enthusiast, the physician Robert Broom, who studied mostly marsupials (Broom, 1895–1915). This comparative morphological tradition has been sustained by Kratzing (1971–1988), also examining marsupials, Wöhrmann-Repenning (1977–1993) on rodents and carnivores, Bhatnagar *et al.* (1976 – on) with bats, and with the first modern monograph, on a prosimian primate (Schilling, 1970).

The chapters that follow deal with the increasing volume of work on a single, but not independent, part of the chemical senses. The evidence presented comes from the lower and higher vertebrates, with the emphasis given to non-domesticated species wherever possible. The informational range now available fortunately extends from ultrastructure and genome mapping to synaptic probes and ethological detail. Points of difference between main and accessory chemoreception are identified, although, where vomeronasal-specific data are lacking, similarities are assumed unless stated. Comparisons with insect chemoreception have not been attempted, although the intricacies of their pheromonal chemicals and signal functions surpass mere mammalian achievements. Nevertheless, insights into chemical signalling by bony fish have led to some comparisons being drawn (Sorenson *et al.*, 1998).

While the mammals predominate in their integration and representation of the sensory world, their noses still tell the brain directly about its chemical space. To explain the workings of accessory olfaction, we need to trace the path of a signal molecule from the moment it leaves its source until a response occurs in the recipient. The events which occur *en route* will determine the effectiveness of the intended communication.

The perception of a stimulus in the brain due to vomeronasal input does not appear to be accompanied by a conscious sensation. The arousal produced by the arrival of "unconscious odours" is nevertheless of considerable importance for the integration of external and internal signalling. The way in which the body recognises an unknown chemical resembles the way in which the immune system sorts out foreign molecules (antigens) from self-molecules. The correct or defence reaction occurs only to the former, much as the nose detects a message molecule; in both cases a specific and appropriate response results. A large proportion of all genes are devoted to these chemical recognition systems. The proportion allocated to vomeronasal function appears relatively small but no less complex than the major systems.

The accessory olfactory complex emerged from a general chemosensory field in fishes through the segregated cell cluster of amphibians, to the distinct sensory apparatus we now see in reptiles and mammals. A general trend revealed by vomeronasal researches is that of a marked decline within arboreal and aquatic animals, which parallels that seen in general olfaction. In the decoding of the sensory grammar of olfaction, some of the basic rules have been unravelled but our current understanding does not yet approach that of sound and sight.

Accessory olfaction contributes to individual and group survival through its influence on the reproductive processes of the diosmic (or double-smelling) groups. Our understanding of its role in the processing of chemosignals depends on the progress of investigations at all levels of analysis. There are many indications that the secondary system employs mechanisms not shared with primary olfaction, reflecting the divergence of its evolutionary history. The various aspects of its workings are presented in an attempt to synthesise the contributions of diverse disciplines and to reveal the nature of an obscure but intriguing sense.

Such an overview inevitably extends into specialised areas of knowledge and experience. I am most indebted to the colleagues who devoted their precious time in providing expertise and critical comment: Drs. E. Grigorieva, M. Halpern, D. Kelly, C. Mucinat-Caretta, L.E.L. Rasmussen, and B. Schaal — I thank them for their efforts. They are, of course, blameless of any inaccuracies and infelicities that remain. I am also most grateful to those who gave illustrations or allowed me sight of unpublished material: I thank Drs. P. Brennan, E.B. Keverne, M. Perret, C.J. Pfeiffer, 'Bets' Rasmussen, A. Schilling, and H. Schultz for their generosity. I am most grateful to Mike Stoddart for his contribution in writing an admirable Foreword.

If this present offering is to be measured against Bacon's aphorism on the quality of books, then it aspires to be mostly swallowed, some of it chewed and digested — and a little of it just tasted.

1 EVOLUTION

"Scent uptake" in Cotylosaurs (from Duvall, 1986).

"When we can classify odours they may prove the basis for a new taxonomy".

J. B. S. Haldane (1954)

1.1 PHYLOGENY

The vertebrates produce, send and detect information which is conveyed by one or more molecular types. Chemical information of biological value (semiochemicals) which partly or wholly activates the accessory olfactory system (AOS) is transferred during intra- and inter-species communication. The compounds involved convey messages of social importance originating from the need to co-ordinate gamete release. It seems quite likely that gradual improvements by selection of semiochemical molecules and their receptors eventually enhanced the reproductive benefits both for the sender and for the receiver (Sorensen, 1996). The dual olfactory systems interpret chemical input to allow the discrimination of odour molecules from one another and to use the information in the construction and perception of olfactory images. The chemosignal output from the sender may represent direct or indirect emissions. The latter case is illustrated in the Heading Fig. (above) showing signal ingestion via a scent-marked object. The signaller in question is a hypothetical mammal-like reptile, shown investigating a scent-marked object (Duval, 1986).

The vomeronasal sense was originally one of the two forms of chemoreceptor cells distributed in the general olfactory chamber of the bony and cartilaginous fishes, and in the Lung-fish (Muller and Marc, 1984; Hansen *et al.*, 1994). The olfactory bulb as the receiving area of the brain for chemosensory information is also a partially specialised structure (Dryer and Graziadei, 1993). By the amphibian stage, the sensory neuroepithelia were further demarcated by placement of the microvillous VN cells in a pouch-like cavity contiguous with the main olfactory sac (c.f. Figs. 2.7 and 2.8). All living amphibians have a common entry/exit duct for both olfactory areas where present. The accessory cavity, while distinct from the main area, is not fully comparable with the VNO as presented in the terrestrial groups. The organ in reptiles developed a connection wholly with the mouth rather than with the nose. Nasal access was regained by the mammals, with or without the retention of an oral opening. The various permutations of oro-nasal apertures in relation to the external nares and their relative positional changes across the vertebrates are summarised in Fig. 2.3(a).

For terrestrial forms, active rather than passive sampling of their chemical space is required. The forked tongue of modern reptiles, and possibly their enigmatic mushroom body (Fig. 2.12) are stimulus-collecting devices. Mammals acquire and deliver odouriferous samples to the VNO by a sequence of active uptake mechanisms, both behavioural and physiological. These almost invariably involve direct nasal and/or tongue contact with the odourant source. Normal respiratory intake presumably also helps to deliver a proportion of signal molecules to the organ, as will chemoinvestigation by sniffing (Youngentor, 1987; Sobel *et al.*, 1999). The importance of stimulus transport from the general nasal epithelium to the VNE is not known but is most likely in mammals with a separation of VNE and N-P duct apertures [Figs. 2.4(a)–(c) and 2.5].

1.2 DISTRIBUTION

In Amphibia, the VNO and AOB are a uniform presence in all groups so far studied, with the exception of the Mud-puppy (*Necturus maculosus*) — one of the Proteid family of salamanders (Fig. 1.1) (Eisthen, 2000). The accessory (vomeronasal) area is least utilised by the more terrestrial forms, the frogs and toads (Anura), although a tree-frog uses a protein pheromonal signal (Sec. 3.3). The tailed forms show considerable accessory elaboration, with seasonal and adaptive alterations (Dawley, 1992 and 1998; Eisthen, 1992). During the metamorphosis, olfactory capacities alter along with the morphogenic sequence from aquatic tadpoles to semi-aquatic or terrestrial adults. The Clawed Toad (*Xenopus*) has received most attention; the results suggest that details of the adjustments required by life-history changes will clarify many aspects of the AOS evolutionary sequence (Meyer, 1996; Hansen, 1998; Iida, 1999; Petti, 1999; Saito, 2000). In particular, they support the association of vomerolfaction with reproductive competence rather than with an adaptive response to terrestrial demands. Breeding larval stages are another likely

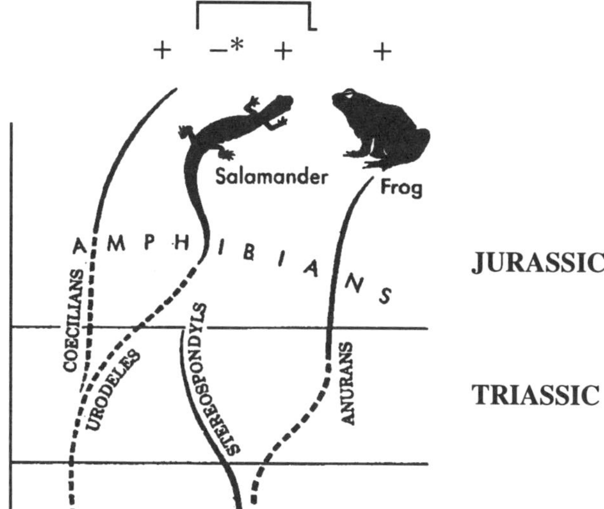

Fig. 1.1 Distribution of AOS in Amphibians. All living groups possess (+) an AOS, except some tailed forms (–*) (after Colbert, 1966).

source of enlightenment; the neotenous Axolotl (*Ambystoma mexicanum*) shows a separate accessory chamber and bulbar region (Eisthen, 1994 and 2000). Neotenous characteristics are not necessarily informative since the avomic mud-puppy also retains larval structures. The aptly named paradoxical frog (*Pseudis paradoxa*) might also repay enquiry. This species reverses the usual trend by undergoing a shrinking metamorphosis, reducing in size by some two-thirds from the tadpole stage.

The generalisation that "ontogeny-recapitulates-phylogeny" could well have relevance for the emergence of vomerolfaction in this transitional group.

1.3 REPTILES

Reptilian accessory olfaction is present in all snakes so far investigated; marine snakes, as secondary swimmers, perversely reduce the MOS while keeping the reproductively more important AOS, as elaborated in their terrestrial ancestors (Gabe, 1976). Among lizards, most are diosmic; even in the burrowing legless forms it is still present e.g. in the Slow-worms (Anguids) and presumably in Worm-lizards (Amphisbaenids) (Bannister, 1968). The adaptive loss in the tree-dwelling Chameleons is associated with prey capture by the anuran-like sticky tongue, since VN structures in the adult are substantially reduced (Haas, 1947). Some arboreal lizards show reduction in AOS structures, extending to complete loss (Pratt, 1948 and 1949; Armstrong, 1953). Snakes and some lizards, while substantially dependent upon vomerolfactory sensation, had to compensate for lack of nasal access to the organ. The morphological disadvantage of using the tongue — a single median structure — as a chemical sampling device to service bilateral sense organs was overcome by the acquisition of forked tips [Figs. 7.3(a) and (b)]. The divided structure neatly fits into the entrances of the VNO ducts, probably not without assistance from ancillary structures (Chaps. 2 and 7).

In the transition towards mammals, the features of a typical intermediate group, the Pelycosaurs (Permian, mammal-like reptiles, Fig. 1.2) probably reflect those now seen in true mammals. The early mammals presumably maintained the oral sampling of odours, as suggested at a

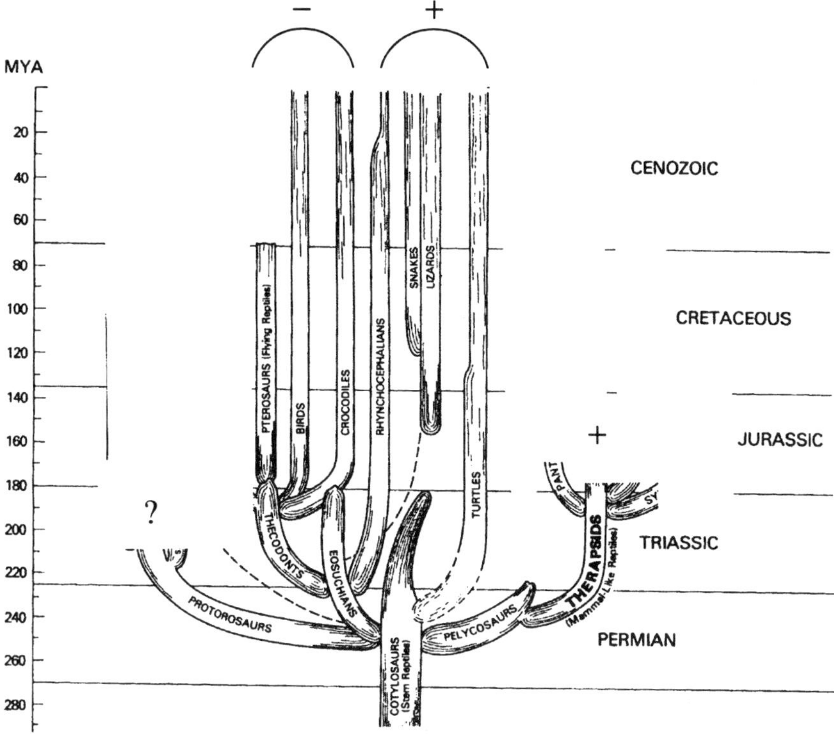

Fig. 1.2 Reptiles: Present in early groups (?) of reptiles and mammals (+), absent in dinosaurs, persistent in only one archaic order (Tuataras, Rhynchocephalia) (after MacLean, 1990).

hypothetical therapsid stage by Duvall (Heading Fig. and 1983). The intermediate steps in this process, such as the appearance of a nasal aperture, will only be satisfactorily deduced given a fossil series preserving sufficient detail of the nasal apparatus. The surviving remnant of these early stages is still known only from the ontogeny of *Sphenodon*. The Tuatara presents a combination of "intermediate" AOS structures. It does not develop a mushroom body, and the VNO opens into both oral and nasal cavities. The embryonic stages are described as having a snake-like dorsal neuroepithelium, in a longitudinal but nasal organ. The aperture is an anterior duct which may not be related to the lachrymal duct (Parsons, 1967 and 1970). The adult condition is undocumented and awaits detailed study. Speculatively, the duct arrangements in the earliest Rhynchocephalians (~200 MYO) presumably resembled those of the earliest "stem" mammals (Fig. 1.2).

The remaining reptiles are monosmic, i.e. they are MOS-dependent with no functional accessory system. They derive from a secondarily aquatic group of Mesozoic dinosaurs, whose survivals are now represented by Crocodiles, Alligators and Caimans (Howes, 1891; Saint Girons, 1976). In these, the loss of accessory olfaction may have been part of a pre-adaptive trend. Genomic comparisons with the avian OR repertoires could provide some clues on AOS history in their living relatives.

1.4 MAMMALS

The presence of the VNO and/or the AOB in the following survey is based upon confirmation by histology, occasionally with histochemistry and sometimes by TEM. A strong behavioural to vomerolfactory linkage allows the Flehmen response (Chap. 7) to be used as a non-structural indicator (Schneider, 1930–1935; Knappe, 1964).

The numbers given below (0/0) show the estimated number of families with AOS present/total family numbers, in those orders so far investigated. The results of a survey of AOB variability generally parallel the pattern of occurrence for the organ (Meisami and Bhatnagar, 1998). The main features of the mammalian organ are evident in the primitive prototherian stock.

Monotremes (2/2): The Spiny Anteater [Fig. 2.12(c)] has a fully differentiated VN complex comprising a cartilaginous surround, vascular and glandular adnexae with a multi-layered neuroepithelium (Broom, 1895, 1897 and 1898). Functional information on the Echidnas and Platypus is lacking; both possess full oro-nasal access.

Marsupials (6/19): These present no known instances of loss or reduction. The Metatheria (Australian and South American marsupials) differ little in the layout of their VN complex from those of the higher placental mammals (Eutherians) (Broom, 1896; Kratzing, 1984; Poran, 1998). Attention to Australasian forms has until recently predominated [Kratzing, 1978, 1982 and 1984(a)–(c); Salamon, 1996]. A few genera of Opossums have been investigated and show extensive control over reproduction by the main and accessory systems. Estrus induction, pubertal onset, early sex-ratio determination and social investigation are among the chemosensory-influenced features (Fadem, 1989; Perret, 1991; Stonerook, 1992; Roland, 1995). Exploitation of the accessibility of the embryonic pouch phase has become an established investigatory tool (Brunjes, 1992; Holmes, 1992).

Eutheria: The minor orders are omitted being of unknown status.

Cetacea (0/10): The Whales are avomic; in both toothed and baleen types other nasal systems (Chap. 2) take over (Aguilar, 1981; Kuznetzov, 1988; Oelschlaeger, 1989).

Sirenia (0/2): In the Sea Cows, the West Indian genus, one of the freshwater Manatees (*Trichecus manatus*) lacks the organ in the adult, and has a reduced MOS (Mackay-Sim *et al.*, 1985). With the marine Dugong, a similar diminution would be expected but is not proven.

Xenartha (1/4): Armadillos are the sole group examined, and in fine-structural detail (de Olmos, 1980; Carmanchahi *et al.*, 1999 and 2000). The Giant Anteater appears to possess some functionality (McAdam and Way, 1967).

Insectivora (4/6): Poorly studied but comparable to other primitive eutherians (Broom, 1915; Hofer, 1982; Stephan, 1982; Matsuzaki *et al.*, 1993); the variety of ecotypes represented, from aquatic to burrowing, are unknown for chemoreceptive adaptations.

Scandentia (1/1): Tree Shrews are relatively well studied for peripheral and central AOS components (Clark, 1924; Loo, 1972; Skeen, 1977; Frahm, 1984). Chemosignals are little known (Stralendorf, 1986).

Dermaptera (1/1): The primitive Colugo/Flying-Lemur, despite its arboreal habits, has a prominent VNO (Bhatnagar and Wible, 1994).

Chiroptera (15/17): The bats (Fig. 1.3) are a heterogeneous group showing considerable variation in diet and in its linkage with chemosensory elaboration (Bhatnagar, 1980). A major subdivision, the Fruit Bats (MegaChiroptera: Pteropodidae), are without a functional VNO or AOB in adults (Maeada, 1992; Bhatnagar, 1980; Cooper and Bhatnagar, 1976). The other sub-division (MicroChiroptera) comprises the remaining families, also with a extensive range of food preferences. Three families have both AOS-present and AOS-absent members, a further three families

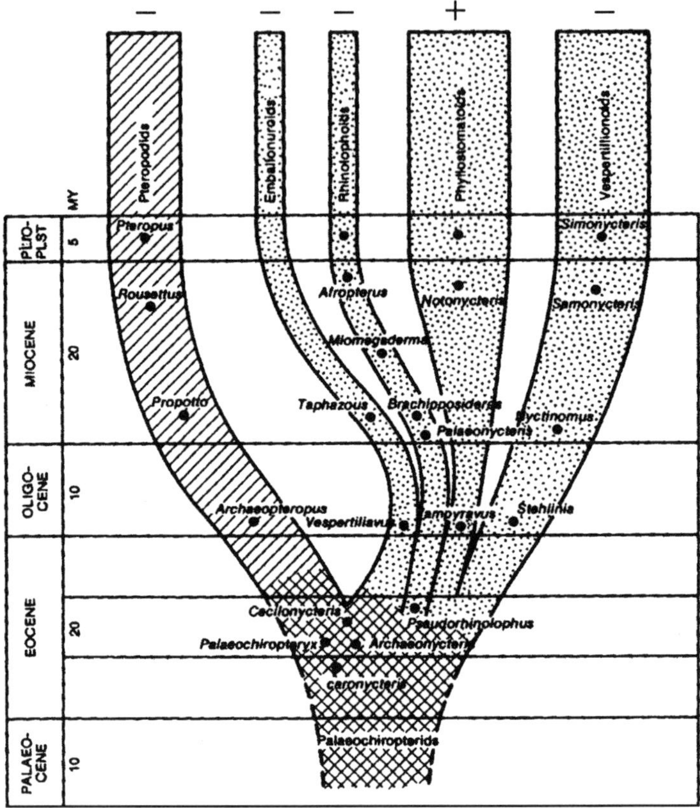

Fig. 1.3 Bats: Uniform absence (–) in Fruit Bats (left: Pteropodidae); uniform presence (+) in Leaf-nosed (Phyllostomatidae); variable (–/+) in other "Small-Bat" families.

which include insectivorous, piscivorous and carnivorous types have no AOS (Frahm and Bhatnagar, 1980; Wible and Bhatnagar, 1996). The structural division seems to lie with families which resemble the pteropodids as fruit or nectar feeders, but which include the sanguivorous Vampire bats (Bhatnagar, 1998). Comparisons of the isolated single genera that are AOS-reduced but within diosmic families such as the Leaf-nosed bats (Phyllostomatidae), would be of interest for functional correlations.

Primates (8/11): This is the other major mixed lineage in which all except the Great Apes and probably the Old World monkeys have a

VNO and an AOB at some developmental stage. All prosimians (Bushbabies, Lemurs and Lorises) plus most of the Platyrrhine (New World) group (the Tamarins, Goeldi's and Cebid monkeys) possess a functional AOS as adults (Maier, 1980 and 1997; Hunter *et al.*, 1984; Evans and Schilling, 1995). Thorough surveys of all genera have not been completed for the New World monkeys, and even less attention given to the apparently monosmic Cercopithecine (Old World) monkeys. The latter have not yet been found to exhibit more than an embryonic VNO (Hendrickx, 1971; Wilson and Hendrickx, 1977). The original loss may well have occurred after the late Eocene divergence of the South American monkeys (Fig. 1.4). The common ancestors of Old World monkeys and of the Hominoids (Apes and Men) possibly included some types with a degree of AOS retention.

The Great Apes are similar to man in the loss of non-sensory structures in the neonate, the more dorsal VNO position on the septum, and in showing minimal post-natal VNE growth (Evans, 1998; Smith *et al.*, 1999 and 2001, in press). The genetic reduction in adult humans suggests a gradual diminution in VN allele number throughout Anthropoid evolution. The absence of discernible VNOR neural elements

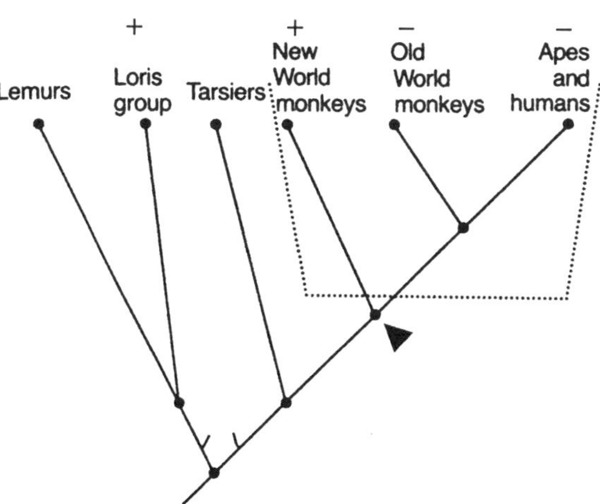

Fig. 1.4 Distribution of VNO in Primates: (+) present, (–) absent (in adults), uniform in Lemurs, Lorises and Tarsiers and non-uniform in Monkeys and Apes. ▲: Progressive (?) loss of AOS capacity (after Martin, 1990).

in adults, together with complete AOB suppression, reinforces the view that functionality is at best atypical, if at all present (Meisami *et al.*, 1998; Trotier *et al.*, 2000).

Carnivora (3/11): A range of structural/functional work suggests that the organ is typically present. Domestic species predominate, e.g. dogs and cats (Read, 1908; Adams, 1984; Salazar *et al.*, 1992 and 1995; Salazar, 1996); the Red fox (Wöhrmann-Reppening, 1993); ferret (Weiler *et al.*, 1999); and mink (Ferron, 1973; Salazar, 1998).

Rodentia (4/30): Of the 2000 or so species, none is suspected of loss. Even a highly fossorial form, the murine Mole Rat, maintains an AOS capacity (Heth, 1995; Zuri, 1998). A promising candidate, the subterranean and eusocial Naked Mole Rat (Batherygidae), shows little or no chemosensory contribution to reproductive regulation (Lacey, 1991; Faulkes *et al.*, 1993). The role of its AOS remains to be discovered, possibly the outbreeding (dispersive) morph is the best candidate type (O'Riain *et al.*, 1996).

Lagomorpha (1/2): The domestic rabbit while similar to rodents, has some distinctive structural features (Chap. 2) (Jacobson, 1813; Wöhrmann-Repenning, 1981; Taniguchi and Mochizuki, 1983; van Groen *et al.*, 1986). The neonate and pups may dispense with AOS chemosignal detection.

Macroscelidea (1/1): The Elephant Shrews as nocturnal scavengers of invertebrates are well endowed for both systems (Broom, 1902; Wöhrmann-Repenning, 1987; Kratzing, 1988).

Proboscidea (1/1): Structural studies are somewhat less frequent than functional (foetal: Eales, 1926; neonate: Rasmussen and Hultgren, 1990; adult: Johnson and Rasmussen, in prep.).

Semiochemical analysis is especially mature in both genera and reveals some unexpected signal sources (Rasmussen *et al.*, 1996; Rasmussen and Schulte, 1998; Riddle *et al.*, 2000).

Ungulata— Artiodactyls (5/9): The even-toed division has one group — the Alcephaline Antelopes showing modifications of the VN and NP ducts, associated with chemoinvestigative alterations (Hart, 1988). Domestic forms again predominate: cattle (Adams, 1986; Salazar *et al.*, 1995 and 1997); pigs (Kratzing, 1980; Wöhrmann-Repenning, 1994); sheep (Balogh, 1861; Kratzing, 1971); camel (Smuts, 1987; Arnautovic *et al.*, 1970); reindeer (Bertmar, 1981).

Perissodactyls (1/3): The odd-toed division is, in equines, without an oro-nasal connection (Jacobson, 1812; Lindsay and Burton, 1984; Taniguchi and Mikami, 1985). No other forms are known structurally; however, Flehmen responses (Chap. 7) in Rhinoceroses predict an AOS presence.

Other than elephants and elephant shrews, a few groups possess modified nasal morphologies which appear unrelated to olfactory acuity. In Elephant Seals and Proboscis Monkeys, enlarged noses occur only in one class of adults — the dominant males, suggesting a limited sex-influenced character. Among ungulates, a bovine, the Saiga Antelope (*S. tartarica*), displays an enlargement of respiratory and protective importance. In Tapirs (*Tapirus*), a feeding proboscis-like nasal extension could contain accessory features as in true Proboscids. Considerable modification to the carnivore rhinarium occurs in one specialised shell-fish feeder: the protective heavy bristles (~400 vibrissae) of the Walrus (*Odobenus*) cover the rhinarium.

The preceding examples may impact upon the chemosensory organisation of one or both systems, but are unstudied.

1.5 MONOSMIC STATUS

Alterations to the AOS raise problems on its role in development and in adult physiology. The partial or complete loss of the VNO implies comparable diminution of the AOB. The bulb is thus proportionately reduced in relation to the extent of reduction in the VN input. The degree of reduction of more central structures is much less well established (Stephan, 1982). Variations in AOB status can give some clues by extrapolation from the least endowed of the diosmic groups (Meisami and Bhatnagar, 1998). Regrettably, the various groups of reptiles and mammals without accessory chemoreception have not been studied in sufficient detail to reveal the intermediate stages and adaptive modifications involved. The AOS functions may have been subsumed by the MOS in whole or part, as in man (Rodriguez *et al.*, 2000). The adult situation in hominoid primates may be extreme and not at all typical of monosmic groups (Chap. 6; Stark, 1960).

The extent of compensatory changes, if any, within the main olfactory system could reveal something of the specificity or generality of

VN input. Complete loss, even if strongly suggested by failure of differentiation in the early embryo, needs to be re-evaluated by a search for diagnostic VN-genome sequences.

Absence from the first developmental stages as in whales and some Old World monkeys suggests that functional substitution is provided by other nasal chemoreceptors. The ganglia and fibres of the terminal nerve system (*N. terminalis*, Fig. 2.9) are the principal candidate (Wirsig and Leonard, 1987). The role of the trigeminal input, although a minor sensor, could well be expanded in a limited capacity (Tucker, 1971; Wysocki and Meredith, 1987; Westhofen, 1987).

A possible alternative is the microvillous part of the chemosensory epithelium in the Organ of Rodolfo-Masera (Septal Organ), which is VN-like (Taniguchi *et al.*, 1993). This mixed receptor population requires much further study since it could prove to have intermediate odourant sensitivities (Giannetti *et al.*, 1995). Its distribution and function(s) are still incompletely known; however it is sufficiently widespread, from Opossums to Rodents, to warrant an intensive survey (Rodolfo-Masera, 1943; Kratzing, 1978; Giannetti *et al.*, 1992 and 1995).

2 FUNCTIONAL MORPHOLOGY

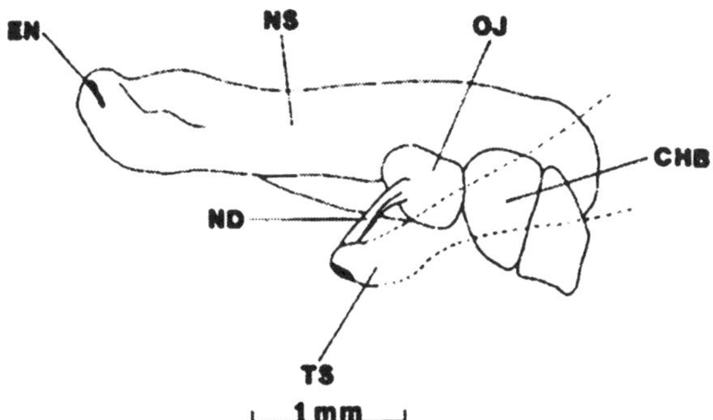

Sensory (vomeronasal) tentacular-apparatus in Caecilians (Apodan amphibian, Gymniophiona) (from Badenhorst, 1978).

"There are those whose study is of smells: how something mixed with something else/makes something worse".

Horace, Book V, Ode3 (translated by R. Kipling)

Ritchie (1944) wrote: "*The Organ of Jacobson consists of paired tubular bodies, vascular and richly innervated, lying enclosed in bone in the front of the nasal chamber, and communicating with the nostrils above, and on the other hand with the mouth by two naso-palatine canals which open behind the posterior incisors. It occurs from Amphibia onwards and is probably an accessory organ of smell*".

This admirably succinct summary made a functional judgement which, although virtually unsupported at the time, has been essentially confirmed. The unique properties of the VNO as a sensory organ derive from its evolutionary history. The adaptive changes relate to the persistence of fluid transport as its principal means of acquiring chemical stimuli. From an evolutionary perspective (Chap. 1), there is considerable support for Broman's view that the VNO is indeed a "water-smelling organ" (Wassergeruchsorgane, Broman, 1920). As such, it was retained as a combined version of "taste + smell" senses in the nose of terrestrial vertebrates, despite the water-to-land transition. Adjustments to chemoreception by obligate air-breathers largely fell upon the main nasal system. The AOS was maintained as a specially demarcated part of the chemosensory spectrum, much as a watery egg is kept by reptiles to provide a protected "closed-box". On land, the primary need was to adapt nasal chemoreception to the dictates of volatile odourants. The AOS component was nevertheless still required to provide detection of some classes of chemosignals delivered via the "aquatic" route. Some of these compounds, although entering the nose as volatiles, nevertheless

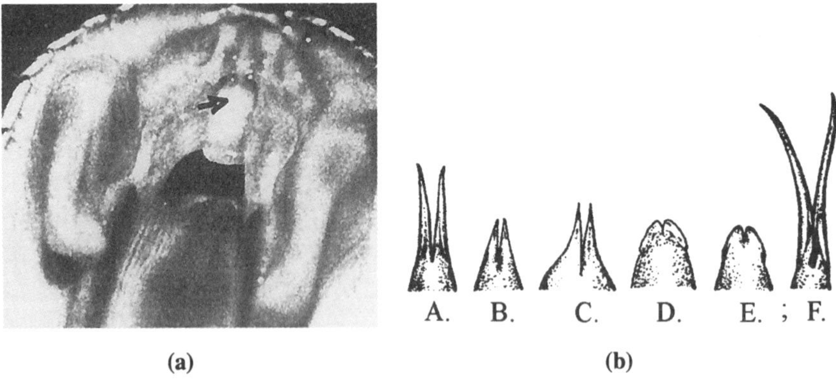

Fig. 2.1 **(a)** Oral aperture of snake AOS: palatal entrance to vomeronasal ducts (VNd) in a crotaline snake (from Miller and Gutzke, 1999). **(b)** Adaptive morphology of reptilian tongue: types of forked-tip in lizards (based on resting tongue length; not to scale). A. Teiid, B. Lacertid, C. Worm-lizard, D. Scink, E. Agamid, and F. Varanid (and most snakes) (after Schwenk, 1994).

could be detected by the VN receptors. The AOS now had to cope with a new medium exerting altered physico-chemical constraints. It also become a short-range detector with limited potential as a distance sense organ.

The higher vertebrates retained the basic *Nervus terminalis*, a VN-associated chemoreceptive neural network (Fig. 2.9) with some cognate functions (Chap. 5 and Murphy, 1998). In mammals, an apparently unique and enigmatic OR cell cluster with a separable MOB input occurs. This minor structure (Fig. 2.10), the Septal Organ (of Masera), is possibly adaptive, has no peripheral connection to the VNO, and its capacity for central modulation of MOS and or AOS input is unknown (Naguro, 1984; Marshall, 1986).

The chemoreceptive systems built up a capacity to discriminate many thousands of chemicals by expanding the range of receptor types. The alterations to the MOS were more varied: from reduction and elimination, to enhancement and expansion. In the AOS genome, the latter trend appears (Chap. 6.4) to be less extensive, perhaps being stabilised to maintain the benefits derived from the linkage between the exocrine and endocrine systems.

2.1 COMPARATIVE ANATOMY

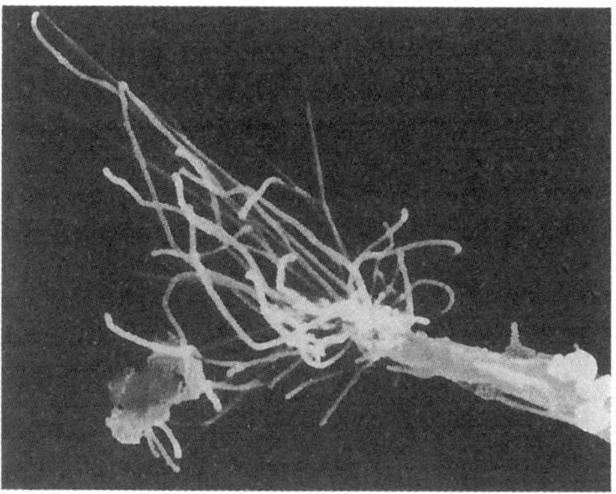

Pl. 2.1A Cellular morphology of neuroepithelium. Microvilli: multiple array at apex of dendrite from an isolated Frog VNOR cell, scale bar = 1 μm SEM (from Trotier *et al.*, 1994).

2.1.1 Lower Vertebrates

The origin of the structural division of chemosensory cells into ciliated and/or microvillous types may lie with the arrangement seen in the present day Protochordates, a pre-vertebrate group. In juveniles of an Amphioxus (*Branchiostoma*), there are primary neurosensory cells with a ciliary + microvillous combination (type I, c.f. Fig. 4.5) and secondary (type II) cells with single or branched microvilli (Lacalli, 1999). These two types are provisionally considered homologues of vertebrate olfactory and taste chemosensors. Sensory cells which conceivably served as mechanoreceptors in the first instance by employing non-motile cilia, then assumed a chemoreceptor role. The emergence of additional sensory cells with microvilli is the presumed next step in chemosensory specialisation. A variety of adjustments of the general morphology of both MOS and AOS [Fig. 2.3(a)] took place amongst main vertebrate radiations (Bertmar, 1981). As mentioned by Ritchie above, the consensus view of accessory chemoreception is that it emerged as a recognisable entity in the transitional (Amphibian) vertebrates.

The jawless and external-parasitic group, the Cyclostomes, comprise lampreys and hagfish, whose simple single-entry nasal chamber contains an olfactory epithelium. Respiratory ventilation by hydraulic exchange through the nares drives water over the chemosensory folds. A set of accessory structures opens below and posterior to the posterior border of the MOE as clusters of vesicles. These diverticula become elaborated during metamorphosis and may respond to "special kinds of olfactory stimuli" (Hagelin and Johnels, 1955). The accessory vesicles are lined by *ciliated* olfactory cells with about 100 times the surface area relative to the MOE (Thornhill, 1972). This device may relate to the slow diffusion of odourants into the vesicles from the main sac.

In the OE of jawed fish only cellular, and little if any, tissue specialisation is achieved. During metamorphosis from tadpole to adult in amphibia, a developmental parallel of water-to-land transition includes the timing of maturation of the AOS. The system as it appears in living amphibians, is already a more or less discrete entity (Fig. 4.3) with its own sub-set of receptors. A process of regionalisation within the bulb, already underway even at the level of organisation in cartilaginous fishes, shows parallel adjustments (Dryer and Graziadei, 1993).

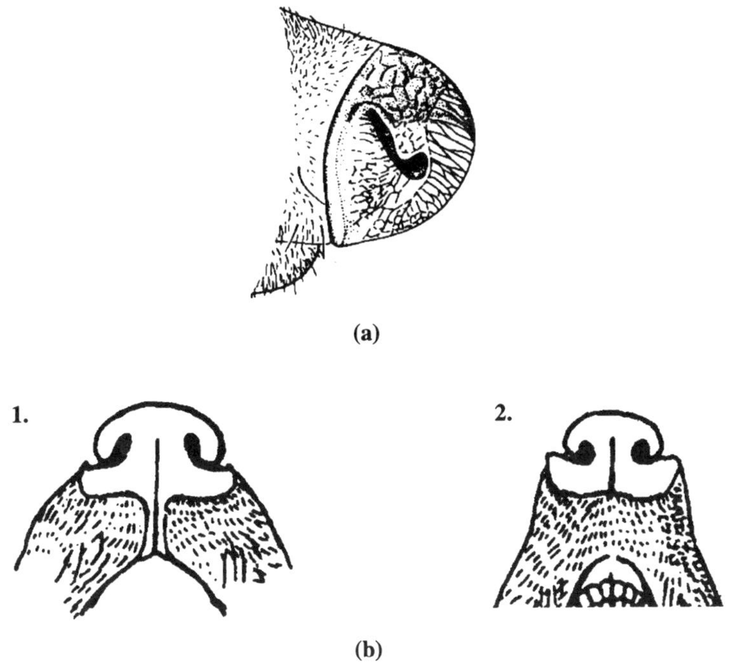

Fig. 2.2 External nose. **(a)** "Rhinoglyphics": lateral view of rhinarial surface, epidermal sculpturing and crenellations in Tree Shrew (*Tupaia* spp.), naked area encloses nostril (from Klauer, 1984). **(b)** Upper lip and rhinarium: variants among carnivores (ventral view, Mongooses). 1. Narrow Philtrum and medial sulcus, 2. medial sulcus, Philtrum absent (from Pocock, 1914).

The variously successful reptile groups, whilst becoming adjusted to the land, nevertheless clung on to the still useful fluid-sensing device which had served their predecessors. The AOS was established as a taste-like capacity distinct from nasal chemoreception, although still functioning under aquatic mode in the early tetrapods. The landward shift during amphibian-to-reptile transition emphasised the more or less functional division already established amongst amphibia. Semiochemicals could now be partitioned by monitoring the respiratory airstream for volatiles, while the accessory system does not rely on nasal intake to deliver its particular chemosignals (Halpern, 1992). The VN receptor cells are stimulated directly by tongue-delivery of odourants, in turn activating the accessory bulb as the first processing centre for onward transmission to higher cortical areas (Inouchi, 1988; Lochman, 1989 and 1993).

The lower vertebrates then possessed a useful nasal device which had already been utilised by their predecessors as a social chemical detector. Air breathing meant that their nasal air-flow was relatively weak in the absence of the pulling power exerted by lungs which

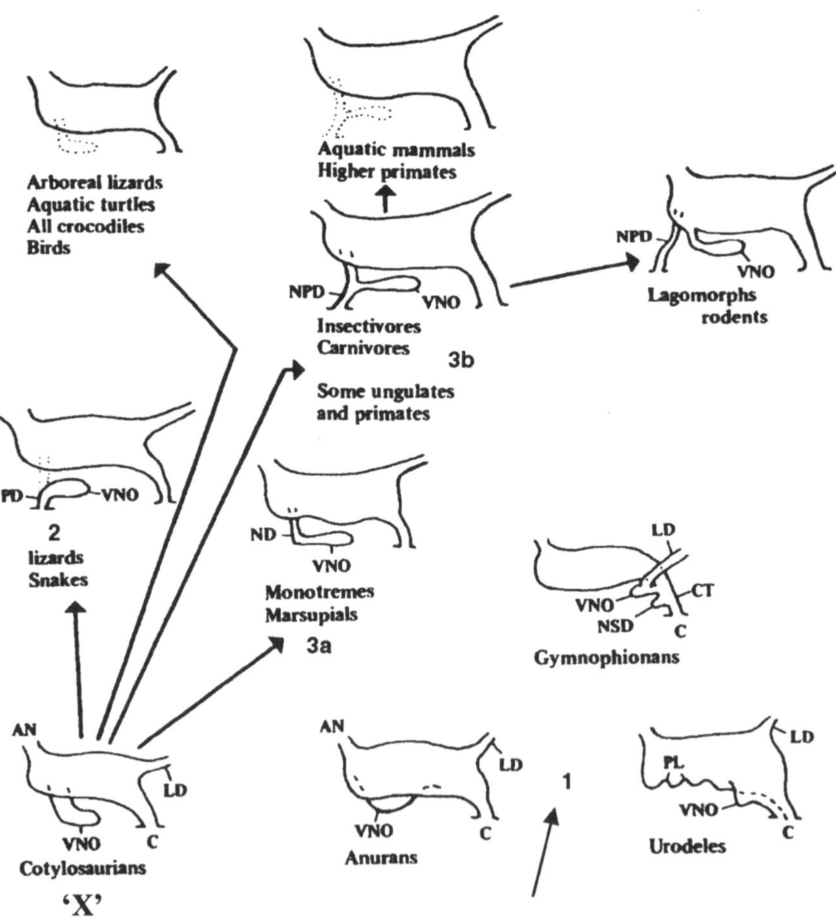

Fig. 2.3(a) Variations in vertebrate oro-nasal duct systems: early Reptile — "X"; Amphibia — 1; Squamates — 2; Lower and Higher Mammals — 3a and b (modified; after Bertmar, 1981).

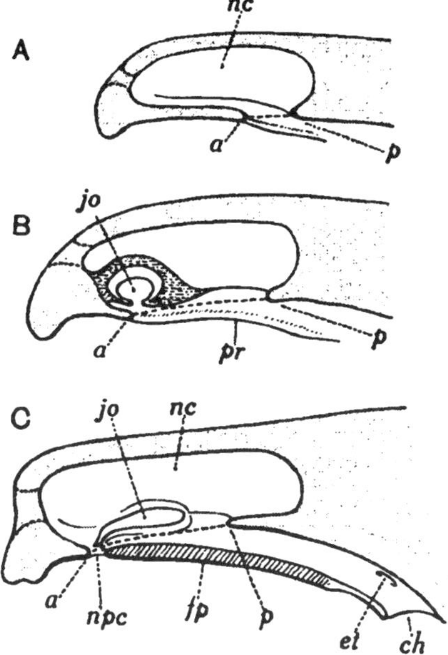

Fig. 2.3(b) Relations of palate to oral and nasal cavities in: A. Amphibia; B. Reptiles; and C. Mammals. a — p limits of internal nostril, ch = choana, fp = false palate, nc = nasal cavity, npc = naso-palatine canal, and pr = palatine ridge (from Goodrich, 1930; after Seydel, 1899).

lack a diaphragm. Some classes of information-bearing mixtures could nevertheless be adequately sampled, provided the problem of intake could be solved.

2.1.1.1 *Fish*

In the fishes, there is a separate nasal cavity, occasionally extended into a ventilation sac, with a linear flow-through current from anterior to posterior. This swimming current can be augmented for olfactory enhancement by grooves channelling into the nares, by ciliary control over L/R alternate flow plus vortex induction across the receptors, and by hydraulic sacs assisting during respiratory movements. A posterior oral access occurs only in the semi-air breathing lungfishes, a remnant family (Dipnoans) with many anatomical peculiarities related to their aestivating habit (Derviot, 1984). The olfactory epithelium is raised into a bilateral series of thin flat plates, termed lamellae, to make a rosette. The arrangement, shape and degree of development of the lamellae vary considerably from species to species. The degree of olfactory development in cartilaginous fish (elasmobranchs) and in the bony fish (teleosts), ranges from the extensive sensory area in macrosmatic species such as sharks and eels, to quite poorly developed in microsmatic visual hunters such as pike (*Esox*), and down to a minimal level in sticklebacks (Bardach and Villars, 1968). Olfactory cell densities reflect the macroscopic picture: increasing from ~10–70 m^2 in sharks and some bony fish up to the ~7.5×10^4 m^2 in eels — the latter being well within the mammalian range (Moulton, 1970; Zeiske, 1989). The Eel (*Anguilla*) improves its acuity/direction-finding by means of moveable intake tubes; incoming flow to the olfactory area can be directed and its rate changed, thus increasing discrimination of concentration gradients.

An unusual type of microvillous cell with a unique combination of dendritic processes has been located by comparative fine-structure studies (Hansen *et al.*, 2000). The Crypt cell has its microvilli accompanied by "sunken" cilia which do not arise from a dendritic knob; further, its surrounding support cells distinguish it from the more conventional receptors. Crypt cells are present in most types of actinopterygians, but absent from sarcopterygians (lungfishes). Whether this apparently specialised feature also has a functional import is not yet established. In the semi-parasitic cyclostomes, there are prominent but unexplored chemosensory abilities in scavengers like the hagfishes and the ectoparasitic lampreys (Braun, 1996). The chemosensory abilities of larval lampreys (ammocetes) are poor, although their superficial similarities to lancelets might have a chemosensory basis.

Sharks are among the largest marine predators, foraging at night, often in deep water, for live and dead prey. Their chemosensory lamellae often show secondary folds bearing ciliated, microvillous, and peculiar "rod-like bearing cells" (Fishelson, 1997). The lateral and medial parts of the olfactory epithelium in sharks and rays project to the comparable regions of the bulb (Dryer and Graziadei, 1993). In bony fish, the olfactory lamellae are arranged bilaterally along a median ridge — the midline raphe — which is a well-defined part of the central lamellar region. The sensory cells can be sequestered in grooves (Fig. 2.14) which may aid the concentration of solutes. The chemosensory content is a mixture of type I ciliary and microvillous neurones (Muller, 1984 and 1998). Some degree of separation of the cells appears on a dorso-medial to ventro-lateral gradation: microvillous > ciliated. Differential projections from the same area of neuroepithelium in channel catfish (*Ictalurus punctatus*) reach discrete posterior bulbar zones, either ventral or dorsal (Morita, 1998). Retrograde fluorescent labelling revealed neurones of two morphological types at differing depths in the lining. Tall ciliated receptors set low in the epithelium project ventrally, while short microvillous cells in the upper part of the epithelium project predominantly to the dorsal region. Hence, although there is the beginning of bulbar organisation in catfish, the morphologically distinct sensory cells remain intermingled across the olfactory epithelium. There seems to be some degree of differentiation amongst the various fish ORs in that there is a generalist olfactory region, whereas responses to more specific pheromone-like compounds are the responsibility of another sub-population of microvillous cells (Hanson, 1998). Whilst in the lobe-finned, relict group, the Coelacanth (*Latimeria chalumnae*) is well-endowed with receptor classes I and II, both receptor cell types co-exist at a similar density in its chemosensory epithelium (Freitag, 1998).

Central/Tertiary structures: The fish olfactory bulb is a four-layered structure much as in higher vertebrates. Within the 2nd layer, the first synapse for olfactory input is on the dendrites of the mitral cells (MC). About 1000 ORN axons converge on one MC, a ratio similar to mammals. The MC output, from cells at various levels, leads into several glomeruli and receives (inhibitory) input from granule cells. The latter also innervate a distinct cell type in the MC layer of teleosts — the ruffed cells (RC), with which they have reciprocal synapses [Fig. 2.18(a)]; both relay cells send ascending fibres to forebrain centres (Kosaka and Hama, 1982). The RC are unlike the MC since they are not stimulated by the ORNs directly. Their interactions (Chap. 5) may contribute to the processing of pheromonal stimuli (Zippel, 2000). The main bulbar pathways project to several nuclei in the forebrain via two ipsilateral tracts, the lateral and medial [Fig. 2.18(b)], the latter mediates sexual behaviour and the former probably other behaviours (Hara,

1994). An extra-bulbar olfactory pathway (EBOP) is present in teleosts and in some non-teleost genera. Olfactory fibres run within the medial forebrain bundle, and can be traced (by SBA lectin binding) beyond the olfactory bulb into areas such as the ventral telencephalon, and/or the preoptic nucleus (Hofmann and Meyer, 1995). The projection of the EBOP fibres is similar in the sturgeon, but in other non-teleosts the primary olfactory fibres reach diencephalic target nuclei.

2.1.1.2 *Amphibia*

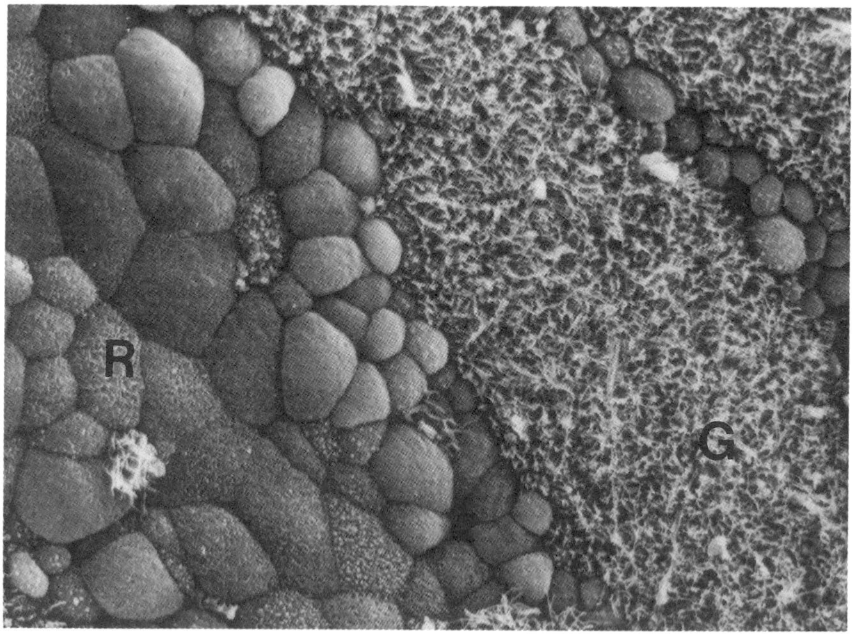

Pl. 2.1B Amphibian AOS, surface of accessory neuroepithelium: chemosensory strips divided by non-sensory ridges in lateral nasal sinus of Japanese Red-bellied Newt (*Cynops pyrrhogaster*); SEM × 742, R = ridge with microvillous cells and G = groove with ciliated cells and cellular protrusions (from Jones *et al.*, 1994).

The first air-breathing vertebrates have continued not only to differentiate their chemoreceptor cell types, but also to show some increase in morphological complexity. Amphibia show several adaptive features for fluid intake on land, such as the tentacles of the Caecilians, shown in the Heading Fig. of this chapter, and the naso-labial grooves of some salamanders (Brown, 1968; Badenhorst, 1978; Dawley, 1989; Himstedt, 1995). Receptor cell size is approximately three times that of other vertebrates at 15 μm, with proportionately sized processes (Pl. 2.1A). There is a columnar arrangement of receptor cells in the edible frog (*R. esculenta*), most of which bear microvilli showing that the process of replacement of ciliated with microvillous cells has continued among the tailess (Anuran) amphibia (Franceschini, 1991; Hansen, 1999). In the wholly aquatic (and neotenous) salamander, the Mexican Axolotl, other organisational features appear, such as distinct clusters of both cell types, with a few support cells bearing motile, current-generating cilia for land-phase smelling (Eisthen, 1994). A highly conserved and diagnostic receptor cell constituent, the olfactory marker protein (OMP), is expressed in all Salamander olfactory neurones (Krishna, 1992).

The location of a "vomeronasal receptor area" is now sequestered in a partially or a completely separate nasal pocket. An early anatomical study described it as "a medial recess of the inferior cavum of the nose" (Gaupp, 1904). In the tailed amphibia, the recess is a lateral sinus (Fig. 2.7), whose sensory pocket is continuous with the main olfactory lining. The appearance of the receptor field, is either uniform or separated into grooves (Pl. 2.1B) by non-sensory ridges (Jones, 1994). The VN area in Common Toads (*Bufo bufo*) is a distinct pouch lying off an inferior recess of the nasal cavity, with an indirect connection to the exterior via a narrow posterior duct linking it to the tubular nasal cavity (Negus, 1958). Separation of primary sensory neurones into two areas gives rise to separate afferent axon bundles. A typical Anuran, the Clawed Toad (*Xenopus*), has clearly demarcated divisions amongst its olfactory neuroepithelia and their pathways. As well as separable main and accessory areas, it retains a fish-like "mixed" middle chamber epithelium (MCE), with ciliated and microvillous receptors (Oikawa, 1998).

In its central projections the vomeronasal pathway, distinguished by a unique lectin-affinity, ascends to an accessory olfactory bulb, while dorsal and ventral pathways supply the dorsal and ventral regions of the main olfactory bulb (Saito and Taniguchi, 2000). The AOS (but not the MOS) of salamanders displays considerable diversity in the pattern of bulbar termination fields, although this degree of variability within a single group was found in a sample of only ten (out of some 400) species (Schmidt *et al.*, 1988). The extent to which such variant structures are adaptive is not known, but it is probably habitat-related. Organisational differences are found amongst particular specialists such as in a wholly aquatic, secondarily adapted, form. Tree-frogs (*Pipa*), have a principal olfactory cavity which retains an organisation into "water" and "air" chemoreceptive regions, as well as the provision of a distinct "accessory" area (Meyer *et al.*, 1996 and 1997).

The two main olfactory tracts [Fig. 2.19(a)] comprise an ipsilateral bundle — in the lateral olfactory tract (LOT) — and some contralateral fibres in the anterior olfactory habenular tract (AOHT). The input from the MOS and AOS shows no inter-familial differences in the final brain locations, apart from minor anatomical variants (Schmidt, 1990).

Analysis of the relative distribution and expression of OR/VNOR genes would clarify the variety of functional chemoreceptive divisions amongst this transitional group.

2.1.1.3 *Reptiles*

The AOS is the predominant sense where there is a complex and elaborated VNO/AOB, as in Ophidians (Lizards and Snakes) (Halpern, 1992). Some chemosignal capacity is used by the Chelonians (Fig. 2.8) but is much less well studied. As noted previously (Chap. 1), the unique Tuatara lizard (*Sphenodon punctataus*) has a mixture of lizard-like and some pre-mammalian features (Broom, 1906; Gans, 1984). Its VNO resembles that of mammals, since it lacks the "mushroom body" (MB) typical of the former. Sphenodon also retains a nasal connection for the VNO, but possibly only as a developmental stage maturing into the normal reptilian oral exit [Figs. 2.12(a) and (b)] (Gabe, 1976). Its adult status will be of considerable interest. In the main reptilian groups, the organ varies from a simple unspecialised part of the main olfactory area found in the Chelonians (Fig. 2.8) to the distinctive ophidian organ. Prey searching and trailing, home-orientation, male courtship and aggressive behaviours (Chap. 7) all depend heavily on AOS input in snakes and lizards. The degree of development often varies with habitat, particularly in lizards, being somewhat reduced in arboreal forms (Armstrong, 1953). Chemoinvestigation and stimulus uptake is almost wholly dependent upon the tongue to bring particulates to the AOS entrance [Fig. 2.1(a)]. The divided tip is usually combined with a narrow tongue, but in burrowing species is combined with a broad fleshy base which also has a demarcated forked tip. The latter ranges from a slight notch, as in skinks, to the narrower and more deeply forked condition of many lizards, as in snakes [Fig. 2.1(b)], being quite distinct in Monitors (*Varanus* spp.).

The MB is a unique structure typical of Ophidians, and forms a non-sensory intrusion whose prominence determines the volume of the VN lumen [Figs. 2.12(a) and (b)]. Its shape produces a crescentic lumen and may act to influence the creation of an inward current, possibly aided by ciliary action. Whether stimulus access to the receptor surface is so improved has yet to be tested. The neuroepithelium is of considerable depth, reaching 0.5 mm in snakes, and is distributed along the curved dorso-medial region. Its complexity is considerable (Fig. 2.15) with multicellular columns within which VNRs and supporting cells are elaborately intertwined (Halpern, 1992; Takami and Hirosawa, 1987). In parallel with the MOS, the relative number of receptor cells varies with the degree of AOS elaboration (Gove, 1979). The differing proportions of receptor cell abundance in vomerolfactory versus main olfactory epithelia range from a high level (Colubrids, Teiids) to intermediate (Skinks) or low (Iguanas). Intrafamilial variation in the extent of MOE and VNE is common, suggesting that there is a comparable variation in the functional role(s) of the two systems (Gabe, 1976). In some groups, secondary loss in the adult would be predicted to follow on from the re-adoption of a watery lifestyle. In the marine sea-snakes (Hydrophylinae), the reduction has occurred within the MOE, now redundant for airborne prey scent. This group may have adopted a compromise derived from their visual-hunting ancestors when they found they had more need of "social" chemoreception via the VNO at the expense of the MOS. Several other correlates of chemosensory function, such as a lingual area taste-bud density, also give some indication of feeding strategies (Cooper, 1997a; Burghardt, 1980). In the Green Anole (*Anolis carolinensis*), an arboreal lizard known to be microsmatic, tongue exploration is infrequent

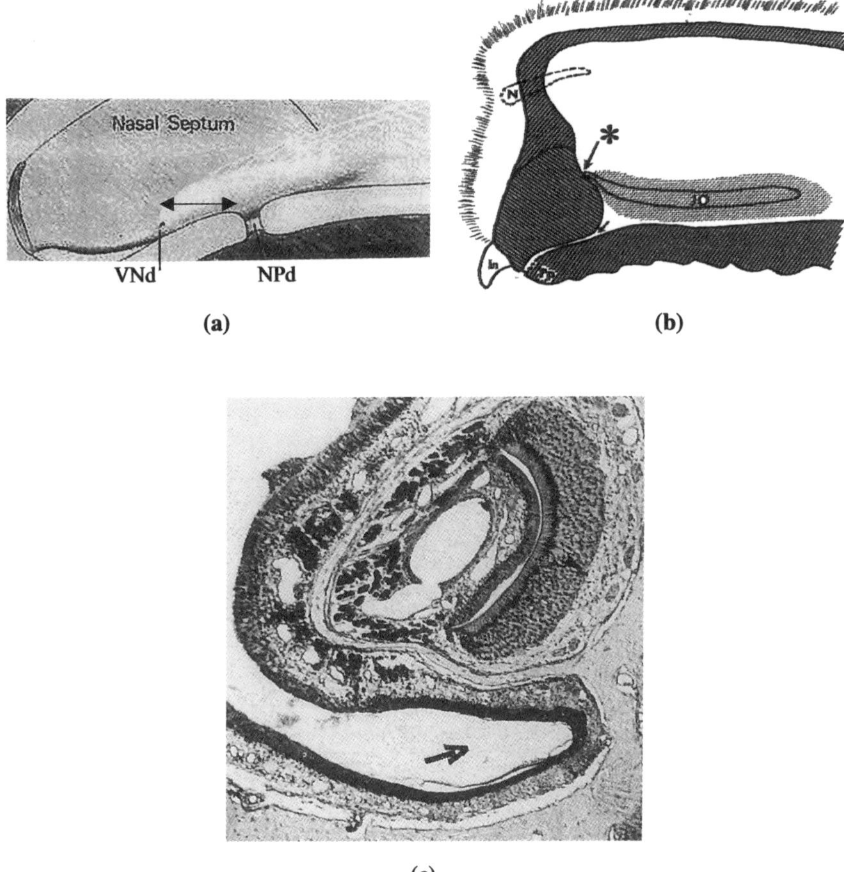

Fig. 2.4 Displacement of the N-P duct caudally and intra-nasal transfer: nasal aperture of the organ (VNd) on the rostral septum in **(a)** rodents, **(b)** * lagomorphs, and **(c)** interconnection of VNd ⟷ to N-Pd by sub-septal gutter in rat (from Wysocki and Meredith, 1991; Wöhrmann-Repenning, 1981a and b).

and has non-chemosensory functions; indeed its minimal VNO is associated with reduced central projections (Greenberg, 1993). Complete adult loss within the Squamates is rare; in specialised hunters such as Chameleons, there is considerable reduction of the AOS from the neonate to the adult (Haas, 1947). This adaptive loss (Chap. 1) is correlated with the "projectile" tongue whose sticky tip allows insect capture, as in frogs.

2.1.2 Mammals

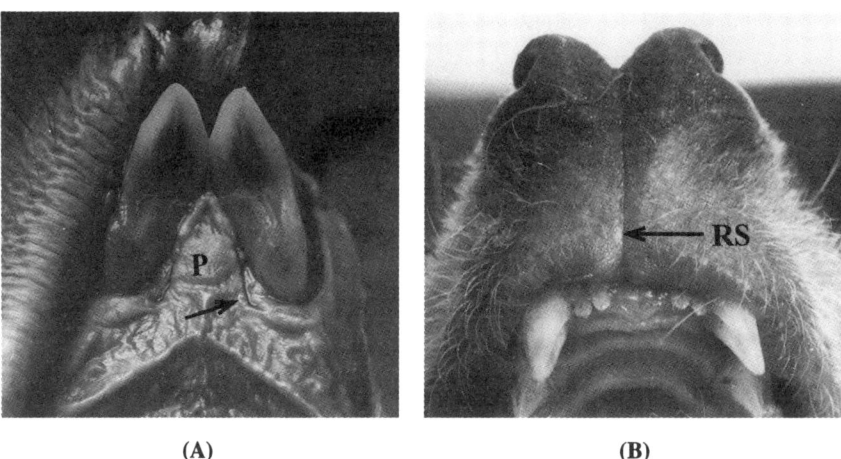

(A) (B)

Pl. 2.2 **(A)** Oral aperture of Prosimian AOS: Naso-Palatine Papilla (Median Sulcus), without incisor-gap, in Aye-Aye (*Daubentonia*). **(B)** Rhinarium and Sulcus, with incisor-gap, in Angwantibo (*Arctocebus*). Arrow → = Papillary Sulcus; P = Naso-Palatine Papilla; and RS = Rhinarial Sulcus (courtesy of Alain Schilling©).

2.1.2.1 *The external nose and ducts*

A clear demarcation of the external nose in marsupials and many eutherian mammals presents as a hairless area of skin surrounding the nostrils and the inter-narial space and can include the median part of the upper lip. This typically pigmented zone — the rhinarium — is conspicuous amongst insectivores, chiroptera, some primates, rodents, carnivores and is often — as in the larger ungulates — maintained as a damp surface by frequent licking. The extent of the naked area is variable, as is the presence/absence of the mid-line groove (the median sulcus). Characterisation of some types of external nose [Fig. 2.2(b)], has suggested a relationship to variables like the shape of the nasal opening and the width of septum (Boyd, 1932). Other associated features include frenulum tethering of the upper lip to the gum, and the highly variable epidermal sculpturing (Hill, 1948; Klauer, 1984). The latter may have some transient role in trapping scents. A broad inter-narial septum and laterally directed nostrils characterise the South American

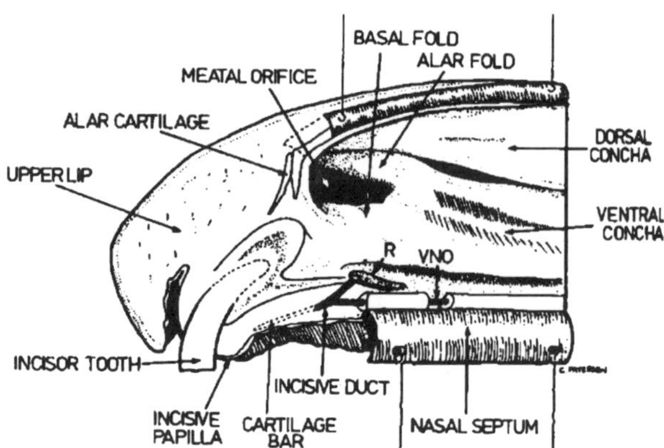

Fig. 2.5 Dissection of horse nasal cavity, with septal opening of VN duct; atrophy of incisive canal (= absence of N-P duct) (from Lindsay and Burton, 1984).

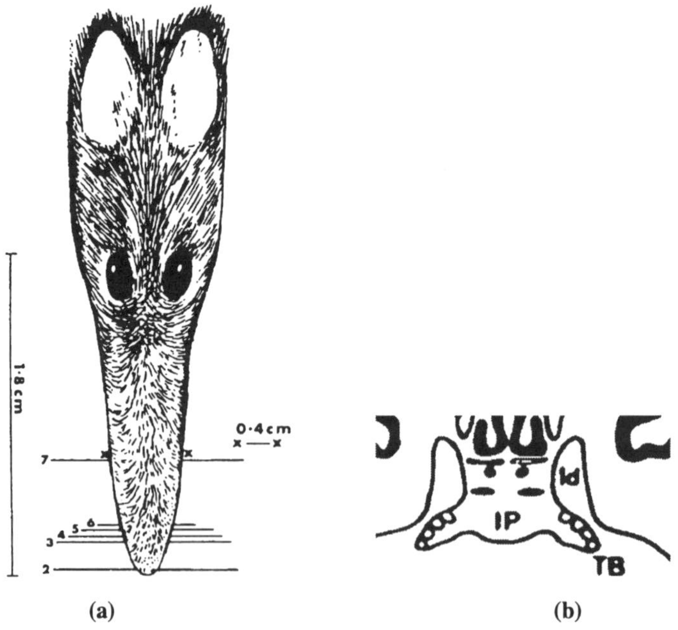

Fig. 2.6 Rostral nasal anatomy of Honey Possum (*Tarsipes rostratus*) showing: **(a)** section levels and **(b)** TS at level 4; naso-palatine papilla with taste-buds (TB), gustatory chemoreceptors facing lumen of N-Pd (incisive duct, Id) (from Kratzing, 1987).

monkeys (Platyrrhines), whereas centrally oriented nostrils separated by a narrow division are found in the African/Asiatic monkeys (Catarrhines). At present (Chap. 1), only the former group (Fig. 1.4) is known to be fully diosmic as adults (Evans and Grigorieva, 1995; Maier, 1997).

On direct contact with scent-bearing material, the various rhinarial crevices possibly have some role in retaining samples until they are removed by the tongue. The median rhinarial fissure or sulcus provides for immediate access and oral transfer of fluids (Schilling, 1970; Prescott, 1977). This groove passes ventrally and divides into a Y-shape just anterior to the median (unpaired) palatine papilla (N-Pp). An associated dental arrangement is the separation of the first incisors by a gap facilitating the passage of the median fissure, as in the Angwantibo (*Arctocebus*, Pl. 2.2B). Specialised feeders such as the Aye-aye (*Daubentonia*, Pl. 2.2A) have an anteriorly placed N-Pp with prominent sulci, but no median access. The papilla extends bilaterally as a flattened structure with flap-like wings whose lateral extensions normally cover the entrances to the naso-palatine canals. When viewed in cross-section, the profile of the N-Pp strongly suggests that in, for example dogs and mouse-lemurs, it functions in some way as a passive "drainage-plug" (Negus, 1958; Schilling, 1970). This may operate after uptake to retain a sample, by tongue pressure or, more speculatively, by vascular changes within the papillary flaps. Mammals without a direct skin route from rhinarium to palatal openings probably sample scents with the tongue as well as relying on the nasal intake. The naso-palatine duct passing

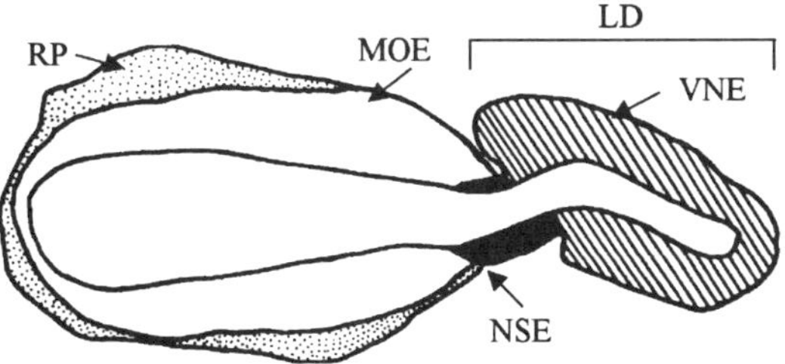

Fig. 2.7 Salamander: nasal cavity TS — anterior to entry of naso-lacrimal duct. LD = lateral diverticulum; NSE = non-sensory epithelium; VNE = vomeronasal epithelium; MOE = olfactory epithelium; and RP = reflective pigment (after Dawley, 1988).

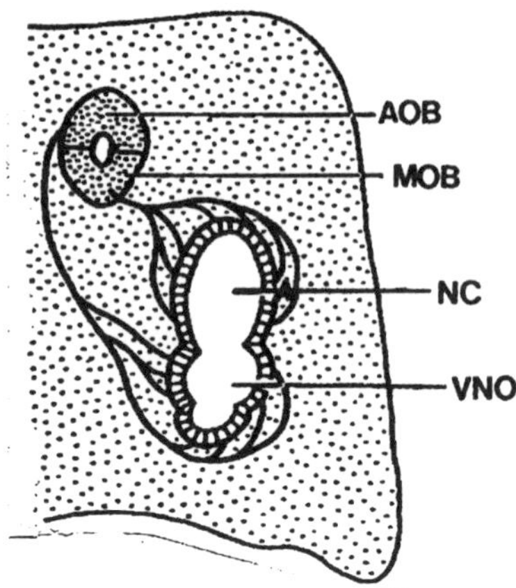

Fig. 2.8 Turtle: schematic TS, accessory structures: lower nasal cavity and dorsal bulb region (after Parsons, 1958).

through the incisive foramen, formed between premaxilla and maxillary bones, is present in most mammals with a functional AOS, and in the majority it connects at its nasal end with the VN duct [Figs. 2.3(b) and 5.6]. The junction with the VN duct can occur at various levels from continuous at the level of the nasal floor, to a median level within the palate, making a modified T-junction with the N-Pd (Schilling, 1970). The nasal openings of the two ducts are separated in rodents and lagomorphs [Figs. 2.4(a) and (b)], the N-Pd being displaced caudally at the level of the diastema, i.e. in the (canine) gap between incisors and pre-molars. Along the nasal floor spanning this gap is a shallow fold or gutter [Fig. 2.4(c)]; its shape suggests that it acts to guide fluids, confirmed by the passage of tracers along it to the VN lumen (Wöhrmann-Repenning, 1980 and 1991; Wysocki *et al.*, 1985).

In genera without a naso-palatine duct, such as horses (Fig. 2.5), the direct nasal opening of the VNd necessarily samples estrous urine during Flehmen by the stallion (Frontisp., and Stahlbaum, 1983; Lindsay and Burton, 1984). The configuration of the nasal folds (Fig. 2.5) suggests that alar and basal ridges may direct incoming stimuli anteriorly and ventrally. As will be discussed later, the expulsion route for lumen contents presumably retraces that of the intake, with fluid ejection via

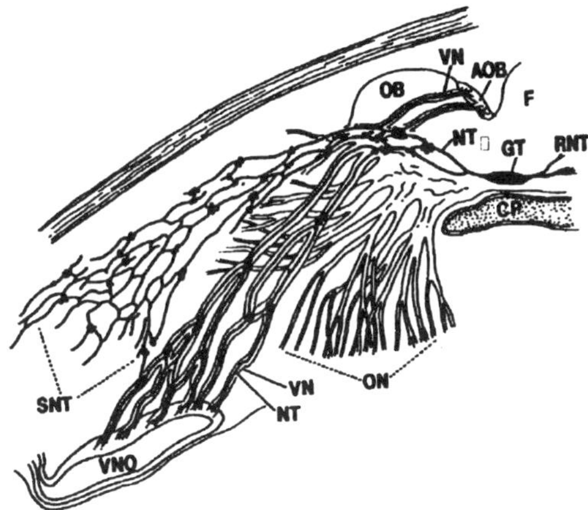

Fig. 2.9 Nasal chemoreceptive systems (Terminalis; MOS and AOS) in neonate Rabbit. CP = cribriform plate; F = forebrain; GT = ganglion terminale; NT(SNT) = *Nervus terminalis*; ON = olfactory nvs.; and F = forebrain (after Huber and Guild, 1913).

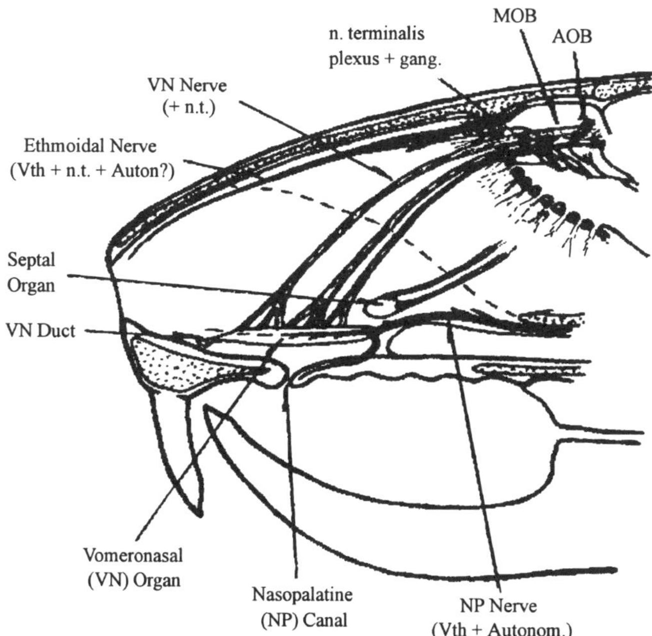

Fig. 2.10 Nasal chemoreceptive systems (Rodent) — chemosensory and autonomic fibres; Masera's organ (= Septal Organ) and NT, and vasomotor (NP and Ethmoidal) in adult hamster (from Meredith, 1983).

the internal meatal opening in evidence during Flehmen (Lindsay and Burton, 1984).

In the squamous/stratified epithelium covering, the palatal aperture of the N-P canals and the dorso-lateral surfaces of the papilla, there are occasional clusters of taste buds. These non-olfactory chemosensory elements are positioned at or near to the entrance to the AOS, suggesting that some initial chemosensation may arise from the sampling of material

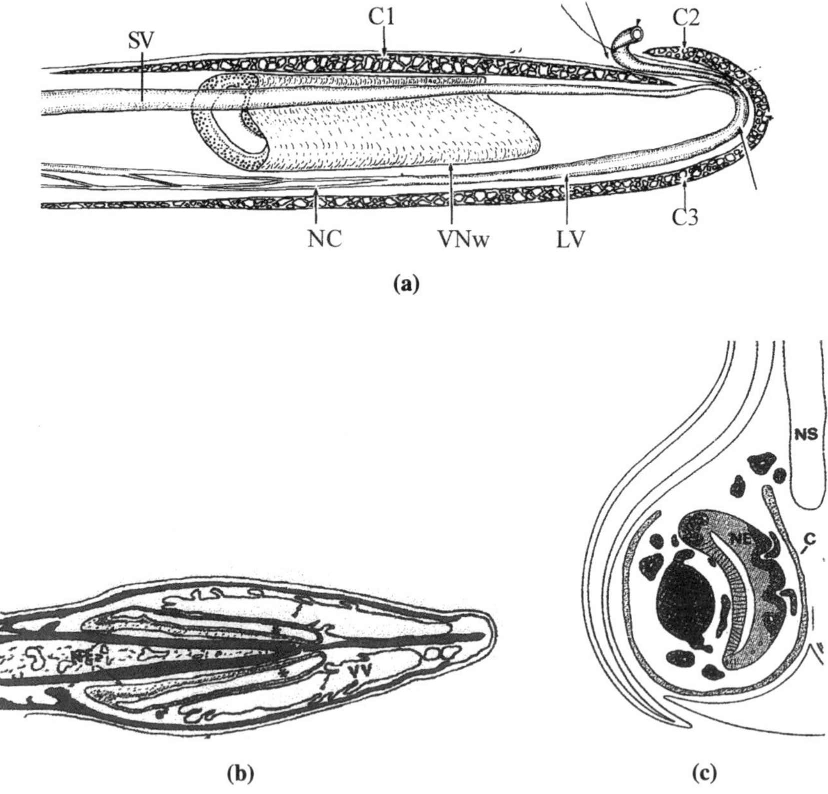

Fig. 2.11 **(a)** Dissection of VNC — Mouse Lemur (*Microcebus murinus*). C1–C3 = Para-septal cartilage bars; LV = ventral vein; NC = arterioles/capillary network; SV = dorsal vein; and VNw = ventral wall (from Schilling, 1970). Vomeronasal complex in murine Rodents. Comparison of LS with TS in Rat: **(b)** LS (horizontal). VV = vascular sinus; arrow = venous diverticulum; * = VN lumen; and NE = neuroepithelium (from Larriva-Sahd, 1994). **(c)** TS (coronal). G = glands; RFE = non-sensory epithelium (from Mendoza, 1993).

passing over the papilla along the sulci. Palatal, non-lingual taste buds are relatively common, but they do not have any assignment to any (receptor) category. Their occurrence is so far limited to minor non-primate groups, the Elephant-Shrews (*Elephantulus* spp.), a Tree-shrew (*Tupaia glis*) and some prosimians and marsupials [Fig. 2.6(b)] (Hofer, 1980a and b; Wöhrmann-Repenning, 1978; Kratzing, 1987 and 1988).

2.1.2.2 *Non-sensory tissues*

The blood supply to the VNO arrives via the ventral branch of the sphenopalatine artery, which divides into both major and minor networks (Salazar, 1998). The arterioles of the latter form a microvascular network supplying the *lamina propria*. The main arterioles of the vomeronasal plexus run medial and ventral to the vomeronasal organ itself, as well as giving off branches alongside the entrance to the VN duct [Evans and Schilling, 1995; Fig. 2.11(a)]. Several small vessels supply the rostral segment, whereas the main one courses along the mid to caudal region. Side-pockets to the vessels [Fig. 2.11(b)], only seen in longitudinal sections, provide some variability in their capacity. The contractile portion of the vascular walls was found to be more developed in the lateral than in the medial wall (Soler, 1998). Such asymmetry presumably helps to exert an inward pressure towards the VN lumen during the expulsion phase of vascular–expansion ↔ lumen–compression [c.f. Figs. 5.7(a) and (b)]. Differences between the musculature of the arteries medial and lateral to the VNO are shown after administration of a transmitter-mimic drug — isoproterenol (an agonist). This produced selective dilation, supporting the view that blood supply to the organ can operate in a similar manner to that of other erectile tissues (Salazar, 1997 and 1998). The vasomotor pump is regulated by autonomic activity; adrenergic (efferent) fibres relay sympathetic stimuli from the Superior Cervical Ganglion routed via the Sphenopalatine Ganglion (Eccles, 1982; Meredith, 1994).

The glands of the nasal cavity which empty their secretion directly into the VN lumen do so typically via one or more ducts which enter at the junction of the two epithelial types, where present [Fig. 2.11(b)].

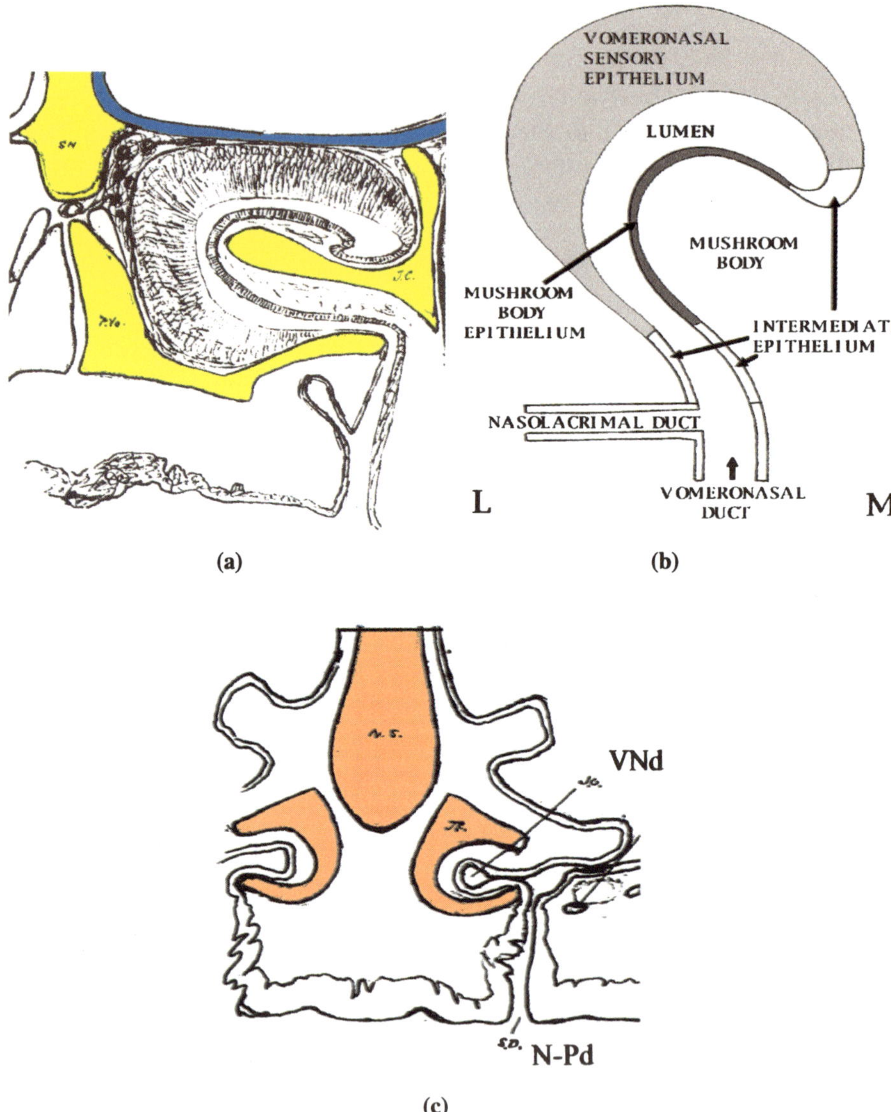

Fig. 2.12 Vomeronasal complex; Reptiles, TS Snake. (**a**) Drawing of *Amphibolurus muricatus*, skeletal elements in yellow (from Broom, 1895). (**b**) Generalised diagram of epithelial types: L = lateral and M = medial (from Rehorek, 1998). (**c**) VNC, Primitive mammal: junction of vomeronasal duct (VNd), with naso-palatine (Stenson's) duct (N-Pd) in Spiny Anteater (Short-nosed Echidna); brown = skeletal elements — septal and para-septal cartilages (from Broom, 1895).

Glandular types cover the compound tubular or tubulo-alveolar construction, commonly lying in the lower septal region and arranged dorsa-laterally or ventro-laterally to the organ (Kratzing, 1984). In an opossum, there is a large posterior (compound) glandular complex whose main duct is continuous with the VN lumen, the junction occurring at the rostral end of the cartilage (Poran, 1998). It is assumed that the lumenal contents of unstimulated or inactive organs are topped up by a low-level trickle of secretion; the presence of long microvilli lining the ducts supports this idea. Compression of acini by the action of bundles of smooth muscle which surround the glandular complexes will sustain secretory flow (Poran, 1998). Other nasal glands opening onto the general septal surface may well contribute to the mucous layer close to the VN entrance. A source of non-vomeronasal secretion found in

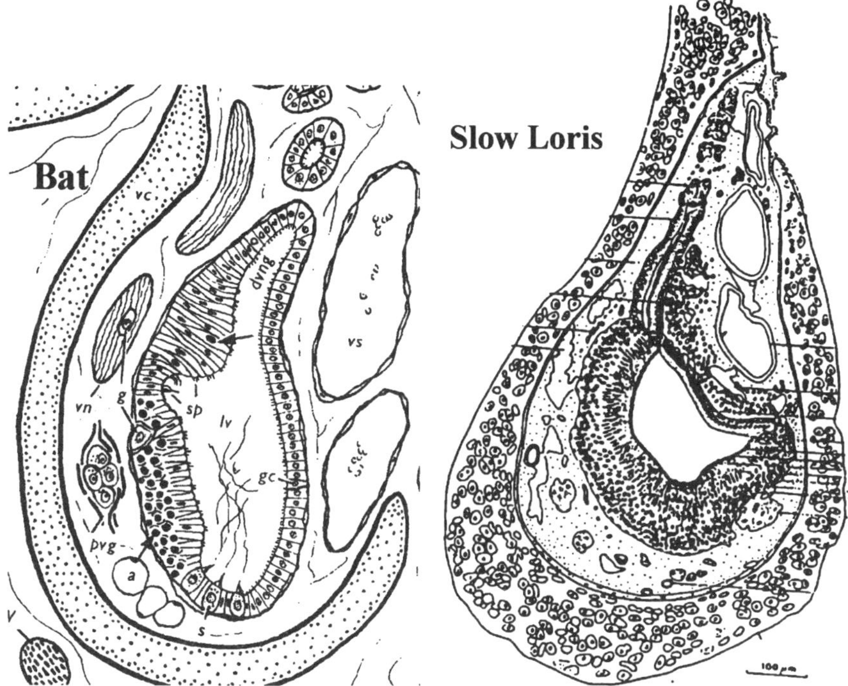

Fig. 2.13 Vomeronasal complex: plan views and diagrams of VNO and adnexae (secretory and vascular). Left: Bat — generalised (from Cooper and Bhatnagar, 1976); arrow = sensory epithelium. Right: Slow Loris (from Hedewig, 1980a); scale bar: 100 μm.

squamates and the Tuatara, but not in turtles, is the conjunctival glands. The lachrymal plus the Harderian glands near the orbit empty into a (common) tear duct, which in turn empties into the VNO duct [Fig. 2.12(b)] close to its aperture (Bellairs, 1950; Rehorek, 1998).

Intra-epithelial innervation is likely to assist in the output of exportable proteins directly from VN glands (Mendoza, 1986). Their secretory cells contain well-developed cellular organelles such as the rough endoplasmic reticulum (RER) and Golgi complexes, related to the presence of secretory granules within the acinar cells. An additional expulsion mechanism is suggested by the occurrence of smooth muscle cells — the myoepithelial content of the basal lamina. Hypolemmal nerve terminals were observed contacting secretory cells, presumably acting to regulate secretion (Carmanchahi *et al.*, 1999).

2.1.2.3 *Neural structures*

The basic layout and functional dichotomy of the accessory and main pathways substantiate the accepted view that the majority of

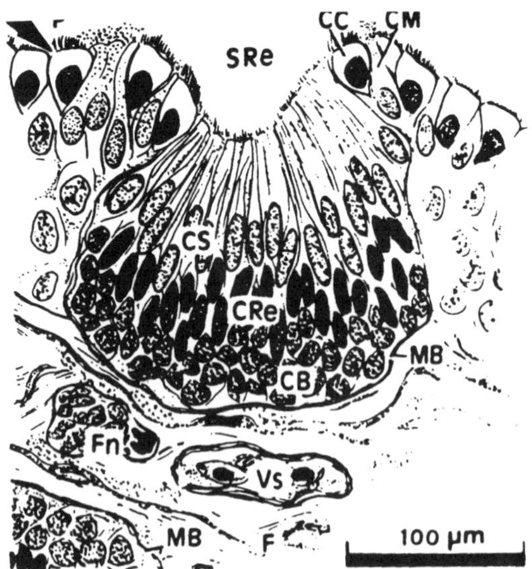

Fig. 2.14 Cellular morphology of neuroepithelium: chemosensory groove in lamellae of the African lungfish *Protopterus annectens* (Dipnoan); basal, receptor and supporting cells (CB, CRe and CS) MB basement membrane. Owen, ×400 (from Derivot, 1984).

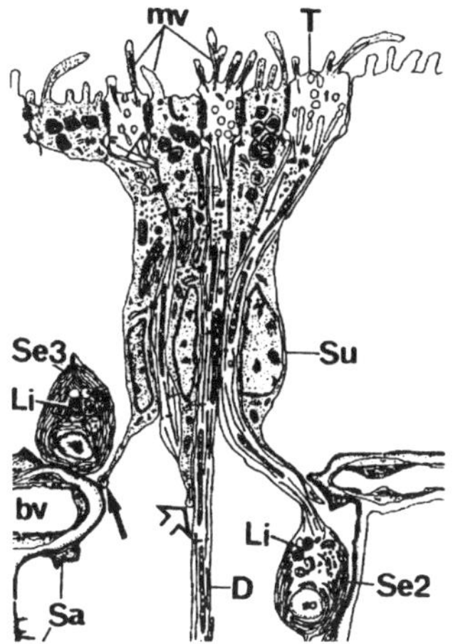

Fig. 2.15 Upper zone of the vomeronasal receptors in a Viper (Habu, *Trimeresurus flavoviridis*) (from Takami and Hirosawa, 1987).

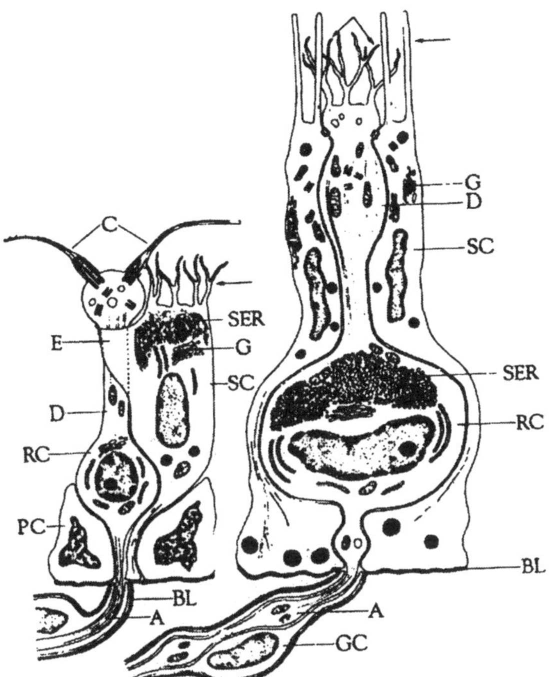

Fig. 2.16 Comparison of the fine structure of VNOR (right) with OR (left) cells. A = axon; BL = basal layer; C = cilia; D = dendrite; G = golgi; GC = glial cell; MV = microvilli, PC = primordial (basal) cell; SER = smooth endoreticulum. TEM, scale bar: 1.0 μm (from Mendoza, 1993).

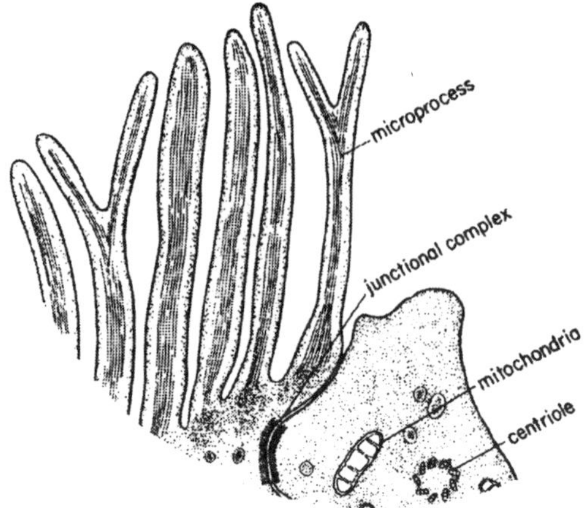

Fig. 2.17 Microvilli in Asian Elephant (*E. maximus*), lumen surface. VNOR with branched and unbranched organelles; tip of supporting cell (from Rasmussen and Hultgren, 1990).

chemoreception is carried out by the two principal olfactory systems (Andres, 1970; Raisman, 1972; Mendoza, 1993). The existence of physiological distinctions strongly supports the evidence from the almost complete anatomical separation (Keverne, 1979 and 1999; Shipley, 1996).

The layout of the mammalian VNC [Figs. 2.13(a) and (b)] is as advanced as the reptilian version but is orally and/or nasally connected in various combinations (Figs. 2.3–2.5). Its neuroepithelium is also as complex, varying in degree of development sometimes with age (Chap. 4) and mostly by group. The VNE extends medially [Fig. 2.11(c)] to circumfentially in some lemurs (Pl. 2.3) and in the "intermediate" primates, the Tarsiers (Wöhrmann-Repenning and Bergmann, 2001). Comparisons of neuronal details (Figs. 2.15 and 2.16) and of the integration of non-sensory structures (vasomotor, vasosecretory and supportive) are left to consideration under physiological functions (Chap. 5).

Output from the bulb in fishes includes contralateral projections in the afferent pathway, while it becomes exclusively unilateral in higher vertebrates, in parallel with that from MOB [Figs. 2.18(a) and (b)]. These centripetal (*inward*) connections of VN and main senses with the

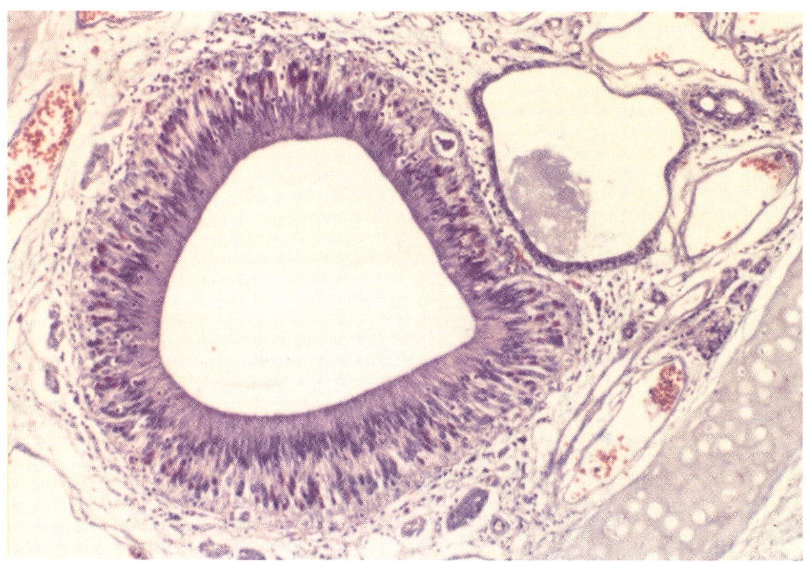

Pl. 2.3 Vomeronasal neuroepithelium in Prosimian: mid-sensory zone in adult *Lemur catta* — uniform lining, ×400 (Evans, unpubl.).

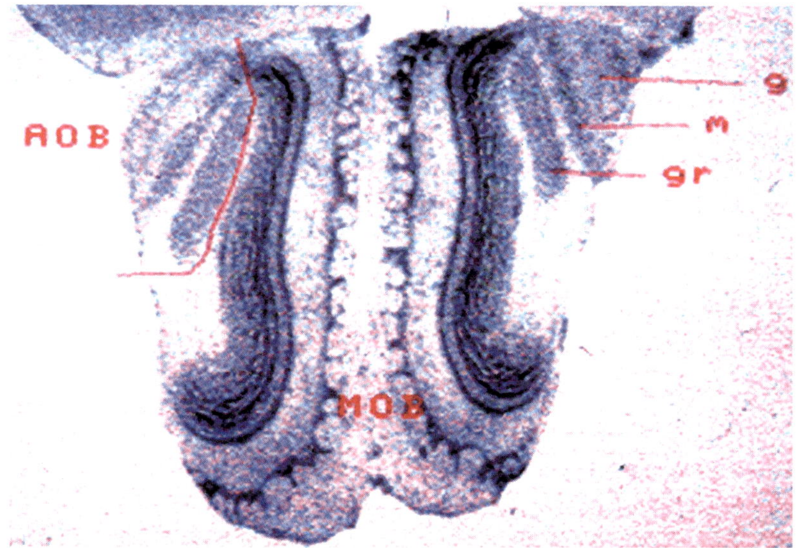

Pl. 2.4 Bulbar layers: horizontal section of MOB + AOB (outlined) in Mouse Lemur (*Microcebus murinus*). g = Glomerular; m = Mitral; and gr = Granular layers (courtesy of Alain Schilling©).

higher cortical centres do not overlap. Indeed, the AOS has a highly restricted projection field in comparison with the MOS. The latter has extensive and widely spread projections (De Olmos, 1978). The amygdala region of the pyriform lobe receives accessory axons in its medial nucleus and in the postero-medial area of the cortical nucleus (PMC); one sub-cortical bundle of fibres passes to the stria terminalis (ST), a feature not characteristic of the MOB efferents. The AOB also receives centrifugal connections [Figs. 2.20(a) and (b)] which return from its central sites as additional input to the bulb (Barber, 1971; Skeen, 1977; Davis *et al.*, 1978). These reciprocal linkages are also a feature of MOS connectivity. Both bulbar areas are then interconnected with sub-cortical structures. Among the most intriguing are those between the limbic system and its sub-cortical regions, the hypothalamus being the most important site.

The internal organisation of the accessory bulbs generally parallels that of the MOB; the arrangement of its cell layers suggesting that it is

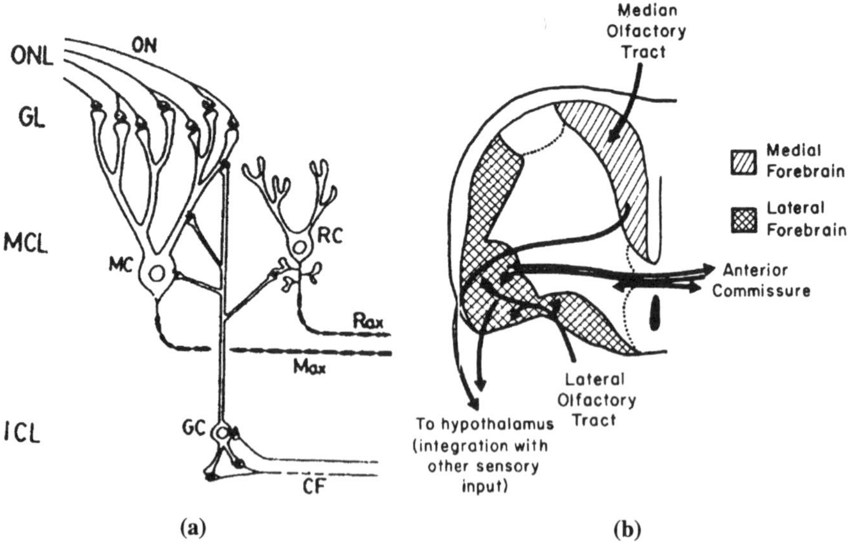

Fig. 2.18 Chemosensory pathways, Bony Fish. (**a**) Peripheral: ON = Olfactory nv. (input axons); GL = glomerular layer; ICL = internal cell layer; MCI = mitral cell layer; RC = ruffed cell; and CF = centrifugal fibre (from Sorensen, 1998). (**b**) CNS interconnections (from Bardach and Villars, 1968).

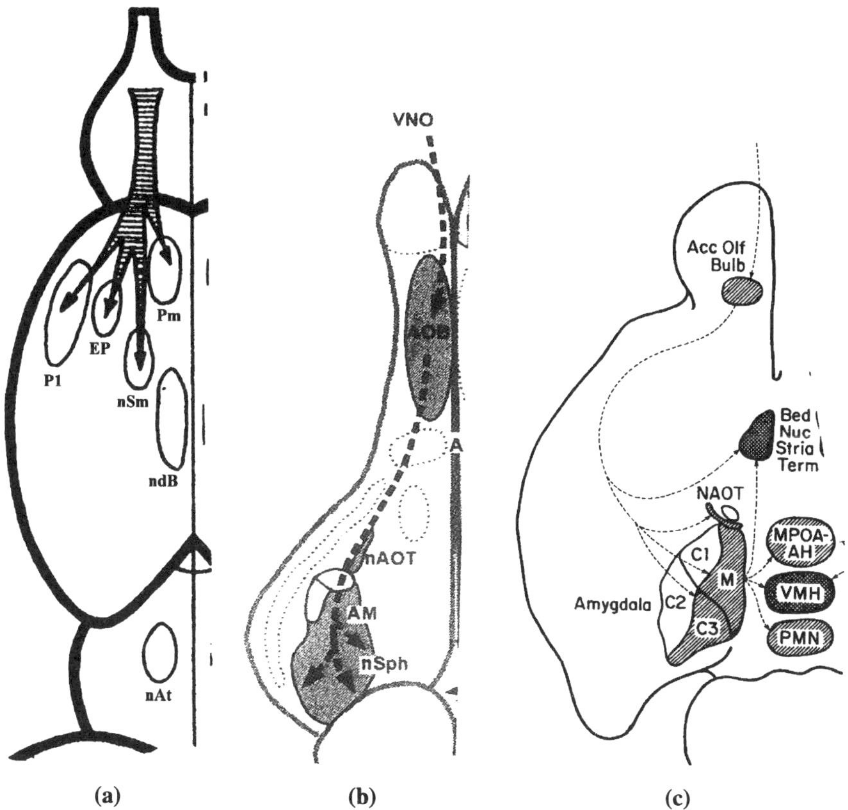

Fig. 2.19 Central pathways and nuclei. (**a**) Frog AOS: Pl/Pm = lateral and medial pallium; EP = post. olfactory eminence; and nSm = medial Septal nucleus (from Kratskin, 1995). Reptiles and mammals — afferent pathways from AOB to amygdala nuclei (Cortical C3 and Medial M), with tertiary connections to other central nuclei in hypothalamus (MPOA, VMH and PMN) (from Johnston, 2000). (**b**) Snake AOS: Second-order projection of accessory fibres; nAOT = nucleus of AOT; AM = anterior amygdala; and nSph = nucleus Sphericus. (**c**) Mammal AOS: Projection sites of vomeronasal fibres in cortex and hypothalamus (from Johnston, 1998).

less highly organised into morpho-functional zones (Pl. 2.4). Secondary connections of the VN axons occur in one or more recipient glomerular zones comprising dendrites of the Mitral plus the Tufted (M/T) cells. The output from the weakly laminated AOB projects into a prominent

Nucleus Sphericus, into the BNST and into the caudal/posterior region of the Medial Amygdala in both reptiles and mammals [Figs. 2.20(b) and (c)]. The fibres run along an ipselateral AOT (the MOS has some contra-lateral fibre tracts) with further connections to Lateral/ Posterior Hypothalamus. In the tertiary centres, however, the AOS and the MOS come together as joint contributors to the chemosensory pathway [Fig. 5.15(b)]. The two types of input arrive at common projection areas within the PMC nucleus of the amygdala (Licht and Meredith, 1985). Any onward transmission from this site can be mutually facilitated, as illustrated in the diagram, or from the output PMC shows

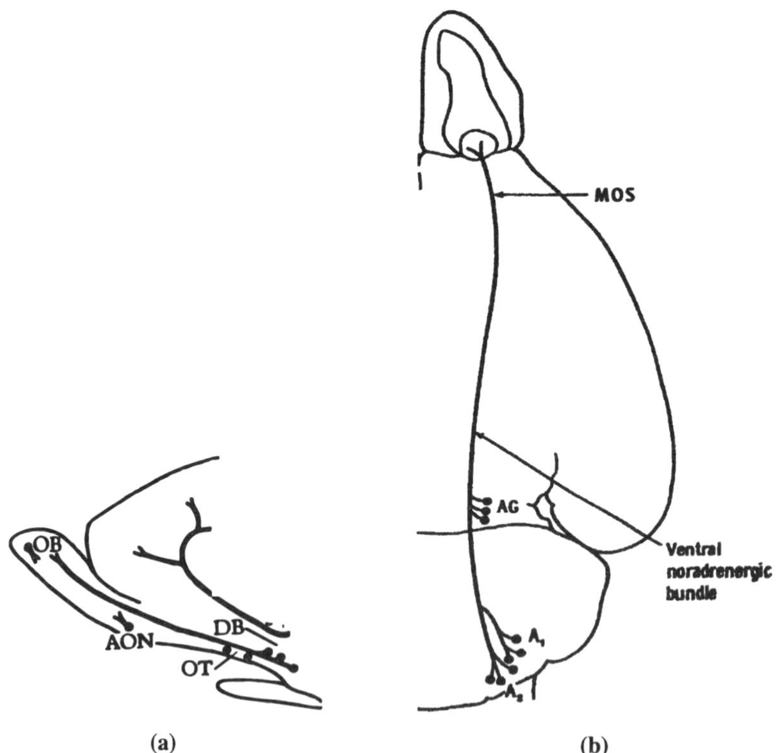

Fig. 2.20 Efferent pathways into bulb: showing (**a**) cholinergic (ACh) fibres projecting to MOB from basal forebrain nuclei. AON = ant. olfactory nucleus, OT = olfactory tract, DB = diagonal band nuc. (from Davis *et al.*, 1978). (**b**) Nor-Adrenalin input to AOB, via MFB pathway from brain stem centres (nuclei A1-2, A6) (from Keverne, 1971).

mutual inhibition. Other units were uniquely stimulated by only one system's input (Chap. 5).

The operation of each part of the vomeronasal sense, from the semiochemicals of frogs to the prevention of pregnancy in mice, is further considered in Chaps. 5 and 7.

3 *CHEMOSIGNALS*

Multiple ChemoSignalling (scent — gland diversity) in Lesser Spotted Genet (from Wemmer, 1977).

"Jacobson's Organ must be a precision olfactory organ which can be excited by a minimal amount of odourants".

Ivar Broman (1920)

The singular and defining characteristic enjoyed by the AOS is the way in which its stimuli arrive. Entry of semiochemicals is by fluid-borne presentation via the nose, nose and mouth, or mouth alone. It is a method which avoids many of the vagaries of entirely air-borne delivery. Chemosignalling by whatever route is an inherently wasteful and inefficient way of communicating information because of the lack of control over the transmission medium. The signal that eventually attaches to a receptor may already have been substantially distorted by non-biological variables. Dilution by rainfall, for instance, could well require compensatory alterations to signalling in many habitats. Increases in glandular secretion volume and/or marking frequency are possible solution strategies. Chemical information arrives at the nose as part of a complex mixture within which the biologically relevant content is distributed, and often amongst very many, apparently redundant, compounds. Some of the latter, particularly the protein fraction, may nevertheless contribute to survival of the signal content, as passive or modulating agents (Hurst, 1998). Downstream from the emission source, the recipient's olfactory systems have to cope with irregular pulses of a mixture whose components become increasingly altered in comparison with their original proportions (Moore *et al.*, 1992). Similarly, air currents will disturb scent plumes of volatiles emitted from scent deposited on a substrate or directly emitted by the signaller (Regier and Goodwin, 1977).

A degree of semiochemical variability is then inevitable during inter-animal transference, offset by some compensatory tolerance built into the system. The long-lasting properties of signal-bearing mixtures, for instance, outweigh the disadvantages of signal variance. Uniquely, odour mixtures show a degree of stability over time which gives chemosignalling its remote characteristics. Unlike calls or movements, scents continue to be broadcast and perceived in the absence of the sender, somewhat resembling the chemical equivalent of a hologram. Overall, an ideal scent-mark mixture should contain a spectrum of volatility from Low → Medium → High, hence providing immediate and rapid dispersal as well as long-lasting sustained release, preferably until the emission can be renewed.

Users of chemocommunication are as likely to be beyond chemosignal range as they are to be out of sound and sight range. If recognisable, signal persistence will maintain an individual's odour-space, provided the maximum concentrations of key signal molecules available are not below the recipient's detection threshold. The pattern of bioactive chemicals should, despite environmental confounds, survive to deliver its message substantially unaltered. Information transfer will be enhanced by any mechanisms that reduce the amount of disruption by the "noise" element of environmental interference between sender and receiver (Chap. 5). The central mechanisms of signal processing have evolved to extract the information contained in variable mixtures. The "encrypted" signal content of the latter is thought to be provided by the relative proportion of several individually meaningless but combinatorially effective compounds, which together convey a high percentage of the intended signal (c.f. Belcher *et al.*, 1986).

Chemosignals are rarely disadvantageous to the sender, although predator urine odours (often sulfurous) repel prey species, while prey scent-trails can betray them to a predator unless they too emit noxious odours (Nolte *et al.*, 1994; Terrick *et al.*, 1995; Cooper, 1997; Chizar, 1999). Other examples of signals with adverse consequences are discussed in Chap. 7. Mutually beneficial cases of plant ↔ animal interaction utilising chemical cues are found in several mammalian orders. Food odours appear to mimic some common social signals since bat-pollinated flowers emit dimethyl disulphide (see below) which attracts *Glossophaga*, a Leaf-nosed and diosmic bat (von Helversen *et al.*, 2000). The bases of similar dependencies among lemurs and marsupials are unanalysed, but could extend the evaluation of chemosignalling mechanisms (Sussman and Raven, 1978).

False or misleading signals appear to be infrequent. The case of female-mimics among garter snakes (so-called "She-Males") who broadcast compounds atypical of males, but not of females, is also dealt with in Chap. 7 on "Behaviour". This type of "deception" cannot be considered the reptilian equivalent of the "Machiavellian" ploys of complex primate societies.

The majority of accessory system usage is carried out in conjunction with the MOS. The combined information flow resulting from dual analysis of chemical input is bound to be advantageous in most circumstances. The advantages of synergistic operation of chemosensory systems are that the MOS detection of volatiles can be followed up by vomeronasal smelling of non-volatiles. Sea otters (*Enhydra lutris*) are reported to locate estrous signals downwind by sea-borne drift, and then to apply genital sniffing and licking at close-range (Riedman and Deutsch: in Estes, 1989). Complementary perception of volatiles and non-volatiles will maximise the effectiveness of chemicals as social information carriers.

Signal complexity presumably favours the operation of a joint analysis of chemosignals. Multiple sources of signal mixtures, such as the local scent-gland array in Genets (Heading Fig. above), is one such mechanism, and others will doubtless emerge (Natynczuk and Macdonald, 1994; Ferkin and Johnston, 1995).

3.1 SEPARATION, IDENTIFICATION AND BIOASSAY

Potential chemosignals are obtained by investigative behaviours, which include nose touching, licking and or close-range sniffing, and sometimes the Flehmen response (see Chap. 7). Chemoinvestigatory responses initiate sensory activity by actively sampling an odour source. Volatile chemosignals taken in with the respiratory airstream will (passively) arrive at the main olfactory mucosa [Fig. 2.1(a) and 2.5]. More usually, odourant uptake will then be maximised by sustained bouts of sniffing during exploratory arousal. An important sexual indicator compound in the hamster is a highly volatile male-attractant, dimethyl disulphide (CH_3–S–S–CH_3), present in vaginal marks and detectable at a threshold, level of $<10^{-12}$ by males (O'Connell *et al.*, 1979). Such "early-warning" compounds serve to induce the close-range chemoinvestigation required by the AOS. Volatiles may aid AOS functions indirectly if they serve to alert and activate sampling mechanisms. Cyclohexanone is seemingly such a "flag" compound contained in the male elephants' secretions from temporal glands. This component is thought to alert female Asian elephants to the onset of male arousal (musth-state). Rasmussen and co-workers found that cyclohexanone **{1.}** elicits chemoinvestigative responses under stringent bioassay conditions (Rasmussen *et al.*, 1998). The signal deemed effective in the opposite direction, by eliciting male arousal to the identifier compound of estrous state, is a multi-taxon pheromone **{9.}**.

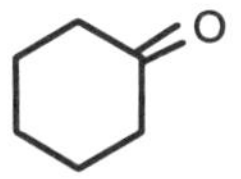

O

Cyclohexanone

{1.}

Regrettably, there is a paucity of measurements on the threshold concentrations that activate the AOS. The fraction of chemosignal input that is directed towards the organ should preferentially excite the AOS, not the MOS. A typical stimulus complex which differentially activates the organ occurs in the snake. A bioassay using earthworm extracts recorded positive responses for prey attacks down to a level of 25 μg/ml (Wattiez *et al.*, 1994). The detection minima for such low- and/or non-volatile signals are poorly known for mammals. One of the best known is a ligand–protein complex contained in the hamster vaginal discharge, in an excess such that the threshold amount (10 μg) judged by bioassay with males, is about one-tenth of the quantity in a single female scent-mark (Singer *et al.*, 1990). Sexual distinctions in semiochemical mixtures can substantially aid bioassay, especially in cases where the compounds are gonadally dependent. Where castrates and females can be distinguished from intact males by exclusive signal combinations, analyses of sexually dimorphic communication should prove revealing. Sexually active males of the Brown Antechinus produced high concentrations of two pyrazine derivatives and four methylketones (Toftegaard *et al.*, 1999). Clear inter-sexual distinctions may characterise the metatherian mammals (c.f. Russell, 1984; Salamon, 1996), but too few representative species from major groups have been examined. Other than metabolic by-products, the main signal sources in reptiles and mammals arise from the array of skin glands whose products derive from their thermoregulatory and/or waterproofing origins. The structure and occurrence of the watery sweat glands and fatty sebaceous glands are extensively documented for reptiles and mammals (Schaefer, 1940; Quay, 1977; Adams, 1980; Mason, 1992). Skin gland products of the ano-genital and perineal zone, plus rectal glands, may be dispersed with or without faeces (Heading Fig., above and Wemmer, 1977). Several compounds which add signal value are incorporated with rabbit faeces (Goodrich and Myktowycz, 1972). The associated behaviours which have evolved to express and distribute the various vehicles (saliva, sweat, sebum, faeces and urine, etc.) have been detailed in the behavioural repertoires of all major groups (see Chap. 7; Brown and Macdonald, 1985; Gorman, 1990; Salamon, 1996). A single quite limited area of the body, as with the multi-signalling posterior of the Genet

Fig. 3.1 Urinary chemosignalling: deposition and broadcast by enurination (auto-marking) on "beard" in male Goat (from Ladewig, 1980).

(Heading Fig. above), can provide social information from multiple sources, each glandular site receiving differential investigation, presumably related to content [Fig. 7.9(b)]. Sebaceous products as well as being rich signal sources may contain ancillary products comparable to the "fixatives" of perfumers, which alter the rate of signal release (Montagna, 1974; Hurst *et al.*, 1998).

Evidence has been accumulating that extracellular metabolic activity of microorganisms, mainly bacteria, occurs within skin glands and on the skin surface (Albone, 1997). Sterile human apocrine secretions do not develop the characteristic axillary odour (Shelley *et al.*, 1953). In the salivary secretions of the boar, transformations of the pheromonal 16-androstene steroids **{2.}** were attributed to the microbial flora (Booth, 1987).

{2.}

Hence, modifications of secreted compounds by bacterial action will provide some components of a signal mixture, possibly thereby altering their binding capacity with putative carrier molecules. The pig (and human) steroids may have sexually dimorphic biological activities but were not found to act via the AOS, even though the associated protein (Table 3.1) resembles other vomeronasal "carriers" (Nixon, 1988; Dorries *et al.*, 1995 and 1997). The most effective stimulus for advancement of cyclicity is contact with a mature boar. This effect is diminished by placing a snout mask on pre-pubertal females (Pearce and Patterson, 1992). Salivary, and hence lipocalin, transfer can be inferred from these results, suggesting that the AOS does have some role in pig neurocrine responses.

Amongst the secretions of specialised exocrine complexes, the ancillary products which act as "sticky" compounds are large, often proteinaceous, molecules. Their primary, secondary and tertiary structures being inherently complex are now seen as ideal informational vehicles — alone or in combination with volatile molecules. Much recent work (Sec. 3.2, below) has identified them as the key components involved in close range transmission, and in intra-nasal peri-receptor events. Proteins are semiochemically implicated when their selective removal or presentation alters responsiveness (Belcher *et al.*, 1990; Mucignat-Caretta *et al.*, 1995).

The production of semiochemicals, other than those arising from specialised exocrine glands, can be at low metabolic cost by exploiting by-products from existing pathways, by the incorporation of waste compounds, or perhaps by acquiring otherwise unusable components. The latter may originate from non-metabolisable molecules in the diet. It is possible that "protective" anti-herbivore compounds from plant tissues may account for the presence of cyanide-containing compounds in one of the larger African bush babies. In *Galago* chest-gland secretions, two out of three major components have cyano-groups (Crewe *et al.*, 1979). Pertinently, another genus of prosimians (*Hapalemur* spp.) can process cyanogenic bamboo foliage as their major dietary source (Glander *et al.*, 1989). Common origins amongst food plants are then as likely as common biochemical pathways; substantial numbers of nitriles occur in at least 2000 species (Legras *et al.*, 1990). The

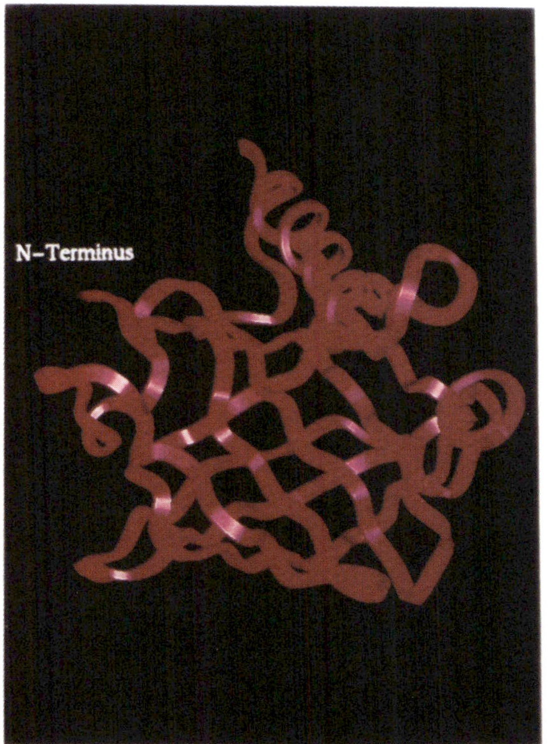

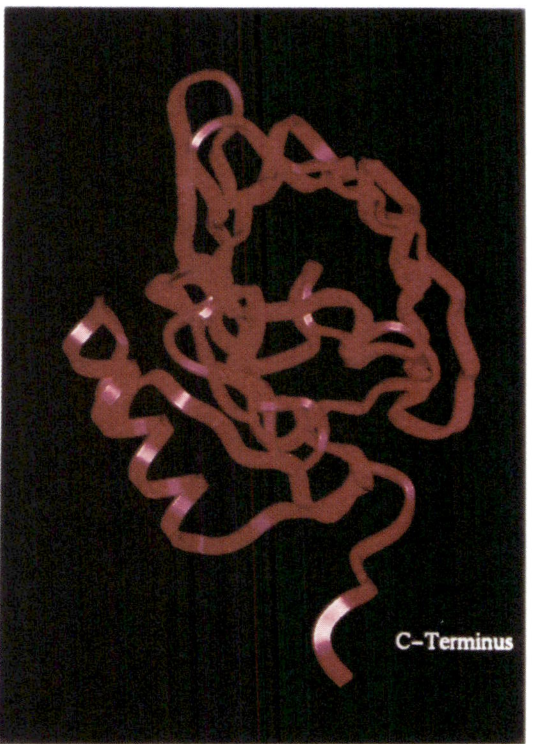

Pl. 3.1A MUP peptide backbone: ribbon representation, shown in two orthogonal projections (left: β-sheet framework; right: binding-cavity, highlighted) (from Lucke *et al.*, 1999).

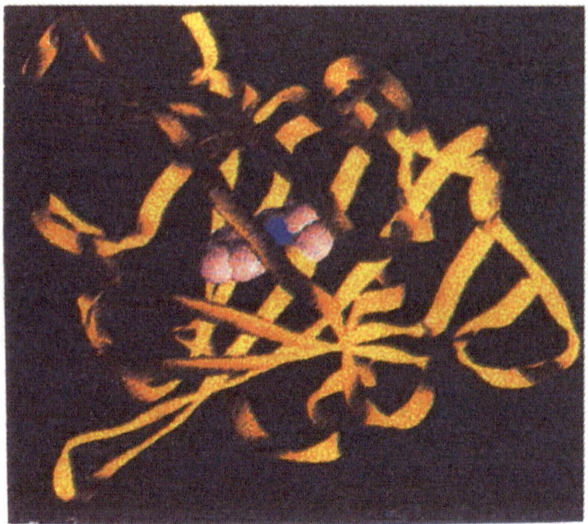

Pl. 3.1B Mouse chemosignal complex, ligand + urinary protein: location of 2-*sec*-butyl-4,5-dihydrothiazole (DHT) within the binding pocket at the N-terminus (lower) in MUP (from Tirindelli *et al.*, 1998).

(E)-2,3-dihydrofarnesol found in elephant temporal gland emissions might have originally been incorporated from the diet as it may in bumble-bees (Goodwin *et al.*, 1999). Whether the appropriate receptor was specifically induced or acquired accidentally by pre-adaptation is currently unresolvable. Even when a "shared" compound such as (Z)-7-dodecenyl acetate **{9.}** is found which does display signal properties, and in spectacularly diverse taxa such as lepidoptera and proboscideans, few useful generalisations have emerged (c.f. Kelly, 1996).

3.2 EXTERNAL TRANSMISSION

There are, amongst the cold-blooded vertebrates, indications of some anatomical and functional segregation within the chemosensory system. Amongst bony fish, the medial olfactory system probably has a separate role in olfaction and has already aggregated some key biosocial functions (Chap. 2 and Dulka, 1993; Kotrschal, 2000). This receptor area largely involves discriminations based upon urinary steroidal and bile components with intra- as well as inter-specific signal functions (Sorensen *et al.*, 1990; Yamazaki, 1990). AOS chemosignals are not apparently designed to act as instant or even "distance" messages due to the factors previously discussed. One of the exceptions representing this class would be the "alarm" emissions of minnows and other related bony fish — "der Schreckstoff" is a pterin released after skin damage (Brown and Godin, 1997). Olfaction in this case appears to out-perform vision by eliciting rapid schooling and/or avoidance: as when samples of water in which a Pike had eaten a roach induced flight in blind minnows (Verheijen and Reutte, 1969). Clearly, alarm pheromones and predator avoidance provide such useful responses that the ability is retained as a generalist signal, acting even between fish species (Brown and Godin, 1997). Such functions persist amongst the terrestrial cold- and warm-blooded groups (Graves and Duvall, 1988; Cocke *et al.*, 1993; Nolte *et al.*, 1994). The inefficiency of scent as a warning, as opposed to the immediacy of alarm calls, can be offset by the avoidance of position disclosure and exposure to predators (c.f. Chap. 7).

Flight by group-living species is likely to improve survival if co-ordinated by mutual chemosensory arousal, with the advantage of producing predator confusion. Whether such examples require a degree of altruistic selection within the social group is unresolved.

With increases in group size and social complexity, and the consequent expansion of diversity in communication, there is presumably selection pressure for unambiguous signals. Over time, evolutionary change may reduce unwanted complexity and act to refine specificity. Whether this results in reliance on the properties of a single component rather than those of the pattern provided by a complex mixture is conjectural. A reasonable expectation is that emitted semiochemical(s) can be reliably produced, and that their contents are acted upon by selection. The improvements in control which can thus be attained are seen in the operation of stereospecificity (Müller-Schwarze *et al.*, 1976).

When the natural product source contains racemic mixtures (of isomeric forms), then clearly the assignment of signal value to either or both variants of a compound needs to be determined. Alterations in receptor detection of chirality can change sensitivity to the geometrically alternate compound over the range 10^1 to 10^6. Of the lactones passed into urine and deposited on tarsal hairs of Black-tailed deer (*Odocoileus hemionus*), only the *cis*-compound with a Z-double bond is behaviourally active at low concentrations. Take-up by licking suggests it may well enter the VNO, as well as reaching the MOE by sniffing. Indeed, this species performs its own online processing of the outgoing signal, by the adsorption of some urinary compounds onto an exocrine lipid film previously deposited on the tarsal hairs. Separation of enantiomers, with retesting of each of the forms, has been elegantly conducted on mouse urinary pheromonal molecules by Novotny and colleagues. They managed to resolve two volatiles with bioactivity **{3.}** and **{4.}**, and then compared artificial with natural mixtures (Novotny *et al.*, 1995).

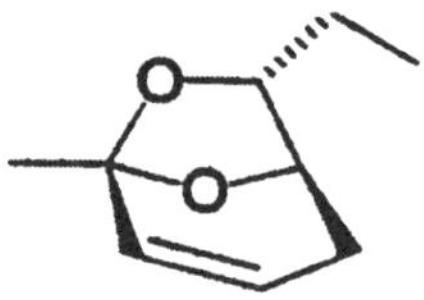

{3.} **DB**

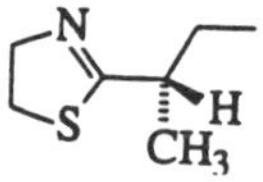

{4.} **BT**

Their conclusions were that the compound **{3.}** 3,4-dehydro-*exo*-brevicomin (DB) was capable of inter- and intra-sexual activity (see Chap. 7), but only if present in the (*R*, *R*)-configuration, while its partner compound **{4.}**, 2-*sec*-butyldihydrothiazole (BT), was probably racemised *in situ*. The DB signal is required in a "correct" stereochemical form but the thiazolines' activity in urine is not affected by a particular chirality (Novotny *et al.*, 1995).

Human axillary odour has biological activity involving steroidal compounds **{2.}**, but it also contains **{5.}** — the noxious (*E*) isomer of 3-methyl-2-hexanoic acid (E-3M2H) — whose threshold is considerably less than that of the (*Z*)-isomer.

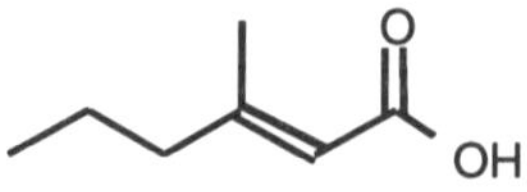

(*E*)-3-Methyl-2-hexenoic acid

{5.}

Although anosmias to these compounds occur at similar levels, some communicative value may arise from the persistence of signal emissions which arc not enantiomerically pure (Carman, 1993; Wysocki *et al.*, 1999). In secreted mixtures, the alternate versions of such compounds are produced in a constant ratio since they havc identical volatility and hence provide stable informational content to the receiver. Support for this idea comes from the results of the NMR mapping of the BT binding site within the MUP1 carrier (Zidek *et al.*, 1999). Here the protein–ligand complex does exist in the expected ratio, and for both enantiomers, although the orientation of the bound thiazole was interpreted as opposite to that indicated by previous X-ray analyses.

Bioassay of alternate molecular forms supports the view that the ORs are capable of resolving isomeric distinctions in neutral (non-biological) odourants. Stereochemical pairs of odours were tested for differential sensitivities in the blind subterranean mole rat (*Spalax ehrenbergi*). The subjects responded to one enantiomer, but not to its stereoisomer. Both sexes were attracted to the odour of R-(−)-carvone but unresponsive to S-(+)-carvone; in contrast, males and females were repelled by the odour of (+)-citronellol, but not by (−)-citronellol (Heth *et al.*, 1992). The lack of responsiveness by mole rats could be central due to lack of salience, or peripheral due to hyposmia/anosmia for one isomer. Both carvones have distinct odours for the human nose.

Other than the internal by-product sources, the main group of signal sources come from the array of skin glands whose products derive from their thermoregulatory/waterproofing origins. Amongst these are large proteinaceous molecules which are inherently complex, and consistently found as components involved in close-range chemosignals. The evolutionary basis for much vertebrate chemosignal design could well derive from early usage employed by amphibia. The current representatives of the order may well retain some/all of their ancestral characteristics in signal composition. Common frogs secrete a large (24 kDa) soluble glycoprotein which is dispersed over their skin surface and which is a potent inducer of attack for their natural predator (Wattiez *et al.*, 1994). The molecule is thought to operate as an identifier (= "prey") via the snake's VN intake. Conversely, when the prey species is a snake, it also uses the VNO to detect traces of its (snake) predator (Miller and

Fig. 3.2 Amphibian protein-chemosignalling: PRF (plethodontid receptivity factor ± interleukin-6), broadcast from male salamander mental gland in the forward current produced by tail-fanning (after Arnold, 1997; Rollman, 1999).

Gutzke, 1999). The AOS of lower vertebrates seems to retain the ability to act as an early warning system in the avoidance of enemies and in the location of prey. This suggests that on land the basis for subsequent elaborations of social chemosignals could have been the simple proteins of immediate survival value.

The production of a female-influencing secretion from the chin gland of male Plethodontid salamander (*P. jordani*) points to a similar extension of function by the acquisition of female olfactory sensitivity to an intercellular signal protein. Female receptivity is enhanced by a male cytokine-like compound of the interleukin-6 family, in its released form. Rollman *et al.* (1999) note that pheromonal activity is a previously unrecognised function for cytokines.

During evolution, minor molecular variants of protein chemosignals have appeared, possibly as contributors to isolating mechanisms which stabilise inter-species differences. Red-bellied Newts (*Cynops pyrrhogaster*) secrete a "solicitation" peptide named Sodefrin (Kikuyama and Toyoda, 1999). In a closely related species *C. ensicauda*, the variant of sodefrin differs from that in *C. pyrrhogaster* by two substitutions: at position 3 and position 8 (of 12). As expected, the male *C. ensicauda* decapeptide is not an attractive signal for the *C. pyrrhogaster* females. Whether such arrangements were present amongst the earliest land vertebrates remains conjectural, but substitution variants could be the simplest mechanism for low-risk/high-benefit signal discrimination.

A sequence of reptilian studies tracked down the likely signal for mating in Canadian Red-sided Garter snakes. Males respond to products on the female's skin surface, which turn out to be related both to insect cuticular lipids and to those of mammalian skin. These integumentary compounds are primary determinants in eliciting sexual behaviour during the "mass-mating" characterising the spring emergence. Up to 100 males surround a female, obtaining pheromonal stimuli by tongue flicking and chin rubbing along her dorsum. The most active extracts of the secretion were the fractions containing saturated methyl ketones, with squalene as an enhancing factor (Mason *et al.*, 1987). The authors speculate that chain-length in these large ketones and/or their double-bond configurations, provide species-specific cues within Garter snakes. Synthetic ketones which most closely resembled natural products were as consistently attractive as the natural products.

Several genera of small mammals use proteins plus small volatile ligands to achieve reproductive co-ordination. Their effect on the AOS has emerged as a primary social function of urine, at least in rodents (Ma *et al.*, 1999). One or more of the five volatile ligands found in mice are designated the causal compounds responsible for a wide range of regulation effects seen under laboratory conditions (Novotny *et al.*, 1999). Each of the liganded signal compounds can bind to the urinary protein (MUP). In consequence, a range of effects on gonadal activity is attributed to the complex.

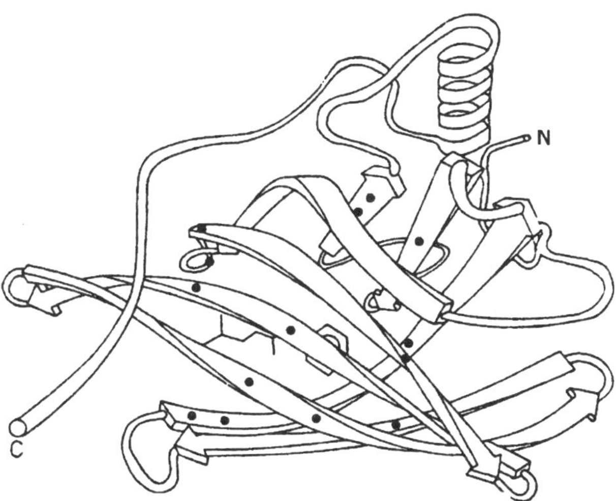

Fig. 3.3 Lipocalins: a generalised 3-D structure showing the β-barrel of eight sheets and the short α-helix (from Schofield, 1988).

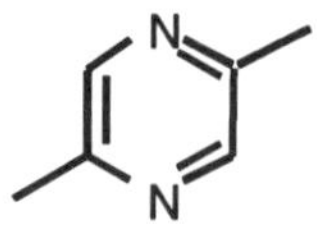

2,5-Dimethylpyrazine

{6.}

2,5-dimethylpyrazine **{6.}** induces suppression of the onset of cyclicity and the maintenance of the inhibition among adult female mice. The advantages of this stress-induced alteration of fertility are that they operate as differential responses to population pressures (Chaps. 5 and 7).

3.3 INTERNAL TRANSMISSION

VN odourants are frequently small volatile molecules combined with large (>300 kDa) molecules. The former may have representatives from a wide range (8+) of chemical groupings (Ma *et al.*, 1999). Once odourant mixtures and VN-specific molecular complexes are delivered to the lumen, they enter the sero-mucous fluid bathing the microvilli. A mix of water-soluble and hydrophobic odourant types will co-exist within the perimicrovillar environment. The current working hypothesis concerning the perireceptor events which precede activational binding of the odourant site is discussed in Chap. 6. These events are assumed, in the absence of contrary evidence, closely to parallel one or more of the models proposed for the MOE (Pelosi, 1994).

To overcome the inherent difficulties of signal movement into proximity with the appropriate receptors, some intermediate mechanism is assumed. The most probable solution is to incorporate a VN-specific molecule of a similar constitution to the binding proteins found in the MOS.

The latter are mostly glycoproteins and belong to a class of 36 kD soluble odourant binding proteins (OBP) which are proposed as local transporters of temporary ligands. The lipocalins are a class of hydrophobic carrier proteins (α2-μ-microglobulins) with common features to OBP in their tertiary (3-D) configuration and a degree of amino acid sequence commonality. A comparison table is given in

Table 3.1 Odourant Binding Proteins: N-terminal sequences of various mammalian urinary and nasal lipocalyins — correspondence/homology with pig salivary protein, pig-SAL (Id <60%) (from Pelosi, 1998).

Protein	Occurrence	Species	N-terminal sequence	Id. %
pig-SAL	Saliva	Pig	**HKEAGQDVVTSNFDASKIAGEWYSILLAS-**	
mus-MUP4	Saliva	Mouse	EEATSKGQNLNVEKINGEWFSILLAS-	46
mus-OBP-III	Nasal mucus	Mouse	EEASSERQNFNVEKINGKWFSILLAS-	46
rab-OBP-III	Nasal mucus	Rabbit	HSHSEVQISGEWYSILLAS-	55
hys-OBP-I	Nasal mucus	Porcupine	EVVRSNNFDPSKLSGKWYSILLAS-	58
mus-MUP	Urine	Mouse	EEASSTGRNFNVEKINGEWHTIILAS-	42
rat-α2-u	Urine	Rat	EEASSTRGNLDVAKLNGDWFSIVVAS-	37

Table 3.1 for the putative carriers in saliva, urine and nasal mucus; the N-terminal sequences are highlighted against the pig salivary protein. All lipocalins or similar uncharacterised (~20 kDa) fractions with biological activity are extracellular, often found in quantity, ~5 g/ml protein in male mouse urine and with a relatively high negative charge.

The rodent urinary α2μ-globulin, as with all members of the lipocalin superfamily, creates a beta-barrel as part of the tertiary folds (Fig. 3.3), which is so constructed that its folded structure produces an interior space containing hydrophobic residues, mostly from aromatic amino acids. Bocskei *et al.* (1992) suggest that successful transport through hydrophilic media may be achieved through the properties of this apolar lining. The configuration of the internal pocket created by the barrel-like "calyx" is a key feature of these dimeric molecules. The existence of some non-selective binding of ligands in the pocket could well be explained by the configuration of binding site(s) within some (e.g. Pig-SAL), but not all similar (e.g. MUP) binding molecules (Bianchet, 1996; Vincent, 2000). The candidate carrier lipocalins so far investigated

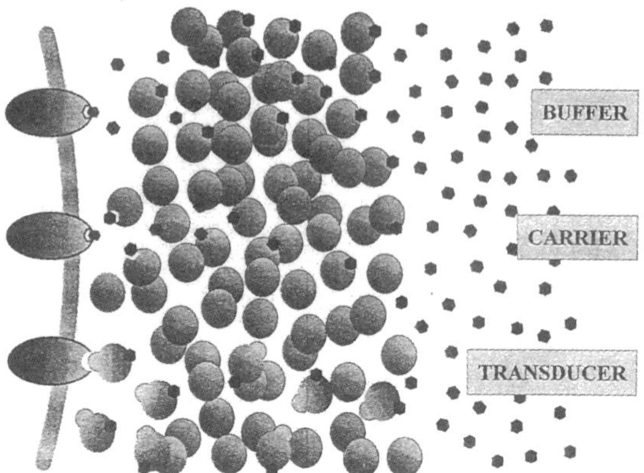

Fig. 3.4 Three models for prospective function(s) for OBPs during perireceptor delivery of a signal molecule: odourant ⇔ protein ⇔ receptor site interactions could involve multiple roles. Ligand ⬣, ± OBP ●, combination(s) buffer and/or carrier and/or transducer (from Pelosi, 1994).

resemble "chaperone" systems and appear to be designed (a) to protect the small volatile partner molecule, (b) to concentrate a rare or low-concentration signal molecule, and possibly (c) to assist its delivery to the VNOR site. Such compounds may then act as necessary co-factors in the transport of odourants, the bound components being chemically hidden in the interstices of the carrier. The DB signal from mouse urine is shown in its ligated position (Pl. 3.1B) at the binding location within the urinary protein (MUP).

A clue as to the importance of signal carriers at both the sending and the receiving stages comes from the occurrence of proteins both as urinary contents as well as in nasal secretions. They are expressed in and secreted by the nasal and other sero-mucous glands. Some manufacture does occur within sebaceous accessory glands; a MUP/α2-microglobulin is predominant in the preputial gland products from the rat (Mancini *et al.*, 1989). Within the VN system itself, a specific carrier compound, another glycoprotein (70 kDa), has been isolated, purified and named Vomeromodulin (Rama Krishna *et al.*, 1994). Subsequently, its carbohydrate linkages were identified (Mechref *et al.*, 1999). Other sources are in tears and from the glands emptying into the trough surrounding the tongue's papillae; a similar lipocalin is secreted close to the chemosensory microvilli (Schmale *et al.*, 1990). There are distinct sites for the production of the OBPs found in each system (Krishna *et al.*, 1994). The location of the main OBP source is the anterior nasal glands, not in the vomeronasal glands; for rat and man the accessory OBP is manufactured within the lateral nasal glands (lng), and in the posterior (vomeronasal) glands. As expected, it is present in all extracellular secretions, i.e. in the mucus covering the sensory epithelia as well as the respiratory surface (Scalfari *et al.*, 1997).

Any one sub-type of the OBPs appears to show affinity for a wide range of similar compounds. This could result from the structure of the odourant-binding site itself, where "new" odourant binding occurs by the displacement of one bound ligand by another (Bianchet, 1996). The general principle proposed is: one odourant molecule for each monomer of carrier (Pl. 3.1B).

Closely related compounds, even enantiomers, may not show similar binding affinities. OBP1 specifically bound 2-{H-3}-isobutyl-3-methoxypyrazine, whereas OBP2 did not; a derivative compound (2-isobutyl-3-methoxy-pyrazine) was unable to displace ligands from OBP-2 (Lobel *et al.*, 1998).

One of the other well investigated protein-based signalling systems arises from the secretions of the genital tract in the female Golden Hamster (*Mesocricetus auratus*). This exudate (female hamster vaginal secretion = FHVS) contains a highly effective "mounting" pheromone (Dieterlen, 1959). Its contents have also been assayed for signal properties, by both behavioural and chemical procedures. Tissue extracts of the vagina and its adjacent tract have shown that there is a high concentration of aphrodisin in vaginal discharge; maximal secretion of the signal carrier arises from the glandular cells found in the lining of the cervix uteri (Kruhoffer *et al.*, 1997). The active fraction was isolated by Singer and colleagues, who confirmed its membership of the lipocalin group of α-2-μglobulins, and named it Aphrodisin. The glands of the uterine and the vaginal linings secrete this soluble OBP, as do the accessory (Bartholin's) and the salivary (Parotid) glands. Its primary structure is that of a 151-AA peptide plus a signal peptide with a 16-AA *N*-terminal (Fig. 3.2) in its secreted sequence form (Henzel *et al.*, 1988). The complementary deoxyribonucleic acid (cDNA) originator sequence shows a high degree of homology (45%) with the cDNA of the binding protein in the rat (Magert *et al.*, 1995). The protein fractions extracted from FHVS yield several related OBPs, including a monomer of non-signal importance (Singer *et al.*, 1989). There is little or no cross-reactivity amongst vomeromodulins, since the mouse urinary proteins do not attract male hamsters and are without any primer effects. This observation is supported by the limited correspondence between hamster Aphrodisin when compared with related rat carriers [e.g. urinary lipocalin (RUP)], since only ten fully conserved positions occur (~6.5%) in their amino acid sequences. Aphrodisin was initially seen as independent of any association with a specific ligand. Indeed, the apparent dual action of the compound is most probably due to its dual nature acting as both "Pheromone and Transducer" when complexed with an odourant (c.f. Singer *et al.*, 1990). There is a characteristic persistence of activity shown by Aphrodisin following dialysis or gel filtration. The carrier–odourant binding is eventually loosened after a combination of filtration plus elution with an organic solvent; the several small ligands still present could not be recombined with Aphrodisin to regain activity. However, in its recombinant form Aphrodisin is capable of binding the volatile semiochemical dimethyl disulfide, as well as other non-signal odourants (Briand *et al.*, 2000).

A source of Aphrodisin which is present in an alternate emission vehicle is that of female saliva. Self-grooming spreads the signal on flank fur and passive or active deposition may then leave traces during contact with burrow walls. The sebaceous flank glands being inconspicuous (and dimorphic) in the female do not play a significant role in advertising her presence.

One of the best-studied "carrier" molecules is produced as a primary excretory constituent of the adult male mouse, known from its consistent high concentration as the major urinary protein (MUP). The basic 3-D structure of the protein was initially obtained from a monoclinic crystal of recombinant protein (MUP-I), constructed by induction in a bacterial expression system and purified to homogeneity (Kuser, 1990). A wild type version of MUP finally yielded to NMR analysis: a clone of the r-isoform (162 residues) was labelled and compared with the crystal-structure (Lucke *et al.*, 1990). Two views of the molecule are shown in Pl. 3.1A, these confirm that the beta-barrel calyx provides an internal pocket which could form the receptive space to contain ligands. The appearance of this carrier-cavity (when isolated) can be derived by modelling of the most likely binding site and is given by Beynon *et al.* (1999). MUP originates from a genc group which also contains codes for a hexapeptide with a common sequence to the N-terminal region of the larger MUP molecule. Mucignat-Caretta *et al.* (1995) suggest that this comprises the binding site recognised by the VNOR membranes. Their conclusion is based on the production of a uterine weight increase in pre-pubertal females by MUP alone. The effect is produced without the presence of bound volatiles and, suggestively, by the hexapeptide alone. The urinary volatile fraction was ineffective by this criterion. The origin of these non-excretory proteins is usually the liver, but only after puberty (Kuhn *et al.*, 1984). At the signal uptake end, however, high levels of the nasal MUP gene variants (4 and 5) are expressed in the pre-pubertal nasal epithelia (Utsumi and Ohno, 1999). The pattern of this expression appears to coincide with that of the OBP I and II genes of rats and mice. If substantiated in other species, the presence of the same carrier at both output and input ends of the semiochemical link confirms the centrality of transport molecules. High specificity during ligand binding suggests that dissociation from the MUP pocket may not necessarily be a pre-activational step in signal transduction (Fig. 3.3). As expected, pure MUP molecules produced by recombinant methods lack any biological activity. Several promising candidate ligand-like molecules have been separated from mouse urine and subjected to bioassay by MUP, that is they can be concentrated in urine by selective binding. Two mouse compounds do show selective binding; the complex acting as the key informational product (Robertson, 1993; Bacchini *et al.*, 1992; Ma *et al.*, 1999). The origin of two estrous-inducing alpha and beta E-farnesenes **{7.a/b}** was traced to the preputial gland in mice.

α β

{7.a/b}

Since these make up ± 80% of the urinary volatiles and elicit several specific behavioural and endocrine responses, they are clearly a major signal source. The range of effects induced by these compounds suggests that farnesenes are likely to act via both the MOS and the AOS.

An alternate source of mouse urinary odour is associated with the histocompatibility complex (MHC); this fraction being composed of a mixture of volatile carboxylic acids. The latter occur in relative concentrations that are characteristic of the individual's MHC odour-type (Singer *et al.*, 1997). Since the mixture of acids varies only in the ratio of the same constituents, as do several other signal mixtures, the likely variant conveying individuality could as well be contributed by a protein–ligand complex. The results from studies of the role of the MHC in relation to chemosensory detection promise insight into mate-selection (Chap. 7). An intriguing possibility is the operation of partner-choice based on non-matching profiles (Wedekind, 1995; Eggert *et al.*, 1998 and 1999). The occurrence of OR gene sequences amongst the MHC sequences may not have functional import, but it might not be entirely coincidental (Fan, 1995).

The efficiency of the putative ligand-carrier binding should be such that it survives rigorous purification/extraction procedures. Of special interest are the characteristics of the boar's pheromonal steroids: 5-α-androst-16-en-3-one **{1.}** and 5-α-androst-16-en-3α-ol. These highly odorous steroids are of gonadal origin, but are selectively concentrated from plasma into the sub-maxillary glands (Booth, 1987). Boars secrete the compounds into saliva where they are kept tightly bound to their carrier, a 20 kDa product originally termed Pheromaxein. The (purified) protein fraction in the eluate retains the steroids attached to a polypeptide sequence containing 29 amino acids, incorporating a sequence (-G-X-W-) identified as "the lipocalin signature" (Marchese *et al.*, 1998). The intact molecule, labelled here as "sex-specific salivary lipocalin" (pig-SAL), comprises 168 amino acid residues with the expected (<60%) sequence similarity to other lipocalins (Table 3.1) (Loebel *et al.*, 2000). The bioactivity of the steroids alone is such that they are celebrated as the first application of a mammalian semiochemical product: "Boar Mate"™ an aerosol, is now routinely sprayed on pre-pubertal female pigs (gilts) as an aid to early mating and improved fertility.

It is in keeping with the complexity of chemocommunication vehicles that alongside the (MUP) complex there exist other categories of, as yet unidentified, semiochemicals. A small pheromonally active fraction in female mouse urine can also produce the well-known male-LH activational effect (Chap. 5), since it appears that protein-depleted urine has as high activity as intact urine (Singer *et al.*, 1988). Some apparently anomalous results inevitably originate in the inherent complexity of the effective fractions. Even where single compounds appear to be the sole bioactive constituent, the purified or synthesised compound does not invariably elicit any VN response. Rat pups, for instance, do not at all resemble rabbits in their nipple location system. The former show rapid orientation to pure dodecyl propionate **{8.}**, normally sourced from their mothers' preputial glands and detected by the AOS (Brouette-Lahlou, 1999).

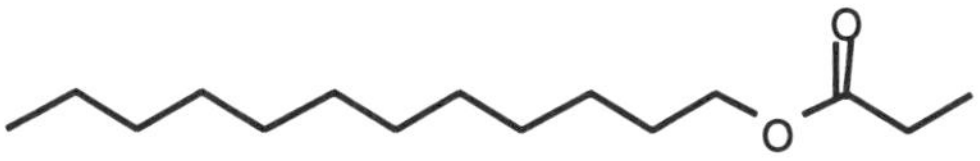

n-Dodecyl propionate

{8.}

In rabbits, the as yet unidentified maternal signal during lactation has analogous properties in guiding the reliable orientation of suckling, mainly via MOS input (Hudson and Distel, 1986; Schaal *et al.*, unpubl.). Minor fractions may still function as "flag" contributors, exemplified by the attractiveness of proestrous elephant urine. Male responses show that intact urine is conspicuously more attractive in comparison with the pure "insect ± mammal" pheromone **{9.}** presented in water (Rasmussen *et al.*, 1996).

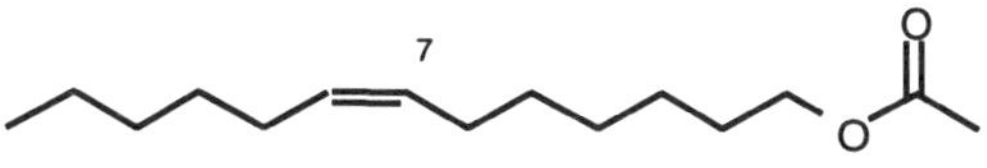

7

(*Z*)-Dodecenyl acetate

{9.}

Boar preputial glands also contain some active compounds not related to the major Pheromaxein component.

Odourant binding proteins have been located in man, adding to the likelihood of some chemocommunication within or between the sexes. Apocrine glands of the human pectoral axillary fields of each sex secrete two similar proteins, one (45 kDa) with similarities to apolipoprotein-J, the other (26 kDa) identified with apolipoprotein-D (Preti *et al.*, 1995; Spielman, 1998). Their importance lies in the ability of the glycoproteins to bind at least one presumptive semiochemical, **{3.}**. This fatty acid has been specifically located (by ISH) in the apocrine axillary glands (Zeng and Spielman, 1996), suggesting that their secretion could incorporate E-3M2H, with the protein content as carrier.

The primary odourant itself is liberated from an initially non-odorous apocrine secretion by axillary microorganisms, these being more active in males (Spielman *et al.*, 1995). The evidence which argues for human axillary-based chemocommunication, is also based on the findings of cycle modulation (Chap. 7) and indirectly on parallels seen with non-primate reproductive alterations. Axillary androstendiones or related compounds have consistently been implicated as potential pheromone-like but not necessarily AOS-active products. Androstadienone is relatively abundant in male axillae and, when presented in air-pulses to the VNO aperture of women, it is reported to be bioactive (Monti-Bloch, 1998). The procedure used is reported to achieve neurocrine modification when a proposed "female" axillary steroid — pregna-4, 20-diene-3,6-dione — is locally delivered on the VN pit of the male human VNO, producing depression of gonadotrophin levels. Serum LH was sufficiently depleted to lower testosterone concentration ($p < 0.01$). However, the afferent route appears unlikely to be that of the usual AOS pathway, since both the VN axonal input and an AOB are lacking in adults (Chap. 5 and Meisami, 1998; Trotier *et al.*, 2000).

Behavioural testing of protein fractions has not kept pace with semiochemical studies. Belcher *et al.* (1990) found that the mixed scent marks of the Saddle-backed Tamarin (*S. fusicollis*) comprise urine and genital/suprapubic gland secretions. Both sexes deposit mixtures of pheromonally active large molecules at, for example, exudate feeding sites gouged into tree bark (Heymann *et al.*, 1989). The deposited complex, once separated, comprises a urinary component of 66 kD and a glandular origin component of 18 kD. These protein fractions have been suggested as likely information carriers (Belcher *et al.*, 1990; Fuchs *et al.*, 1991). The alteration of the signal content(s) by enzymic degradation does not remove all biological activity. Digestion of male scent marks is sufficient to prevent preferential chemoinvestigation, whereas female scent still retains activity. The suggested interpretation is that proteins are a prominent part of the signal, but not the major identifier.

Assessment of estrous state by bulls requires some nine of 20 compounds of various types present in cows' cervico-vaginal mucus to induce full investigation (Klemm *et al.*, 1987). Here, the pheromonal source is remote from the emission/signalling site. The attractive fraction appears in water extracts of vulval skin-gland products (Rivard and Klemm, 1989). Positive chemoinvestigation was elicited by serum from the same female which probably contained both volatile and non-volatile (unidentified) compounds. Nasal contact by the bull induced increased salivary discharge and other indications of autonomic activity. Feedback arousal of the cow is an additional source of increased stimulus output; localised adrenalin produces prolonged perineal sweating (Blazquez *et al.*, 1988) — possibly biasing a bull's choice of female. Other chemosignalling products present in urine and/or cervical mucus may mediate inter-female coordination of receptivity and exert some puberty accelerating influence (Izard and Vandenbergh, 1982a and b; Hradecky, 1989).

The prophecy on semiochemical systematics in the headline quotation at the start of this chapter remains just that — an intriguing speculation. Some single-component chemosignals do turn up as apparently the main active compound in a complex secretion. In male gerbils (*Meriones*) one volatile, phenylacetic acid, appears to represent the "dominant male" state (Thiesen, 1974). Individuality must be added by further components — dietary or variable sebum constituents in this case. Indeed, amongst mammals and some reptiles, complex mixtures seem to be the norm; very few taxonomically relevant examples have emerged.

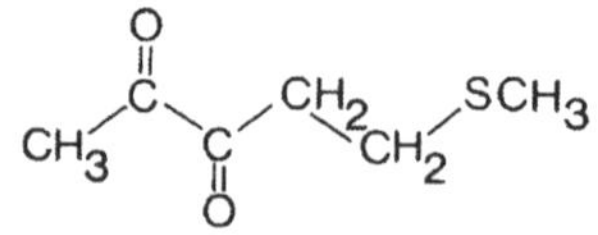

{10.} 5-Thiomethylpentane-2,3-dione

An instance of the problems in chemotaxonomy occurs amongst the Hyenas. A potential marker component **{10}** occurs within anal gland secretions from two species, the Brown and Striped (*Hyaena* spp.), but not from a third — the Spotted Hyena (*C. crocuta*) (Buglass *et al.*, 1990). However, it is a major component only in the Striped Hyena, while five other compounds (out of 20+) are common to all three species. Which signals contribute to intra-species and which to inter-species distinctiveness cannot be disentangled without elaborate and time-consuming manipulations.

Field and laboratory bioassay of chemosignals from related sympatric and allopatric species (overlapping and discrete distributions) are essential to an understanding of the relatedness or otherwise of functionally active compounds. The semiochemicals involved in speciation surely utilise the main and vomeronasal senses, but their relative contributions cannot be predicted at present.

Chemosignals which activate the AOS have been consistently shown to comprise specific lipocalins (VNPr) and/or one of a number of volatile, hydrophobic ligands (L). Activity of the signal complex (VNPr + L) must be retained, whether transfer is immediate by licking of genitalia or postponed when emitted by marking of the substrate. On entry to the peri-vomeronasal mucus (c.f. Figs. 2.16 and 2.17), the complex may disassociate to re-attach to a second-stage but related carrier. Alternatively, it may persist as a protein-complexed signal until the ligand can become attached to the receptor binding site (VNOR). The alternatives are diagrammed in Fig. 3.3, but no definitive role has been fully substantiated.

A sequence of related VNPr carriers has been found to be involved, in the transportation process of the signal for estrous arousal in male elephants (Lazar *et al.*, 2000).

A tentative schema based on these findings is projected for a sequence of OBP and ligand transfer-interactions, from urinary deposit to VN lumen; the series of intermediate steps in Z.7-(12)Ac transfer is summarised as follows:

Female urine [estrous signal(s)] ⇒ Male flehmen ⇒ mucus [OBP] uptake

VNPr-U ⇒ VNPr-T ⇒ VNPr-D ⇒ VN lumen ⇒ VNOR ≈ Z.7-(12)Ac.

(urine) → (trunk) → (duct)

The first step uses a 66 kDa albumin-like molecule as carrier (VNPr-U), from which the acidic conditions of the trunk mucus (VNPr-D) will detach the urinary ligands. The carriers in the trunk contents are additional OBPs which then preferentially bind the lipophylic ligands (L) released from their urinary vehicle. Once deposited at the VN-duct entrance, a third transporter (VNPr-D) completes the final movement into the organ (Rasmussen, pers. comm., 2001).

The train of events from sender to receiver is seen as following the sequence: C* + L ⇒ L (transmission) ⇒ L + C* * ⇒ L~VNOR. At the last step, the G-protein linkage induces the second messenger cascade whose amplification triggers the appropriate ion-channel (Chaps. 5 and 7). The role of the carrier(s) at either end of the transmission sequence is gradually emerging, but without any consensus on their exact contribution, or even on whether they are a crucial component for information transfer (Tegoni *et al.*, 2000).

General rules governing the chemocommunication processes so far identified will hopefully emerge from studies of all vertebrate AOS channels. The design solutions arrived at by the amphibians, which are solely protein-based in several instances, may have persisted unmodified. They may also have been subject to selective change because of the necessities of fluid-based messages in terrestrial animals. The addition of small ligands would at least raise the combinatorial pool of signalling compounds.

A recurrent refrain in the semiochemical literature is the lack of sustained and comprehensive studies covering all transmission stages: from stereochemical detail to ethological nuances (Albone, 1984 and 1997).

4 DEVELOPMENT

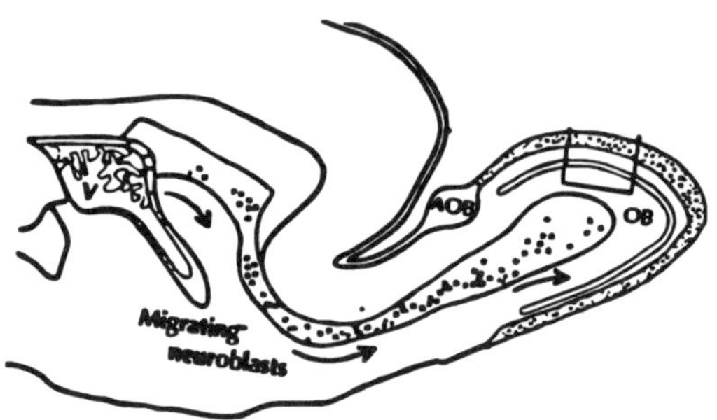

Continuous neurogenesis: downward (centrifugal) movement of neuronal cells into the main bulb (from Calof, 1995).

"Rhinology — '...where in the nose was the heart?'"

Candida McWilliam (1988)

The developmental processes which produce the AOS parallel to those of the MOS and, although they become separate systems as soon as the organisation of the forebrain begins, do not differ except in the details specific to each system. The differentiation of cell types within the organ, synapse formation in the AOB and the establishment of more central tertiary connections occur in a similar but not identical sequence. In general, the differences amount to alterations in the timing of ontogenetic events, with primary olfaction usually in advance of the sequence in accessory tissues. The distinctive morphology of the VN complex does require that the non-sensory elements be produced in co-ordination with its neurosensory maturation. These ancillary but functionally necessary tissues (e.g. ducts) can affect the operation of otherwise mature sensory capabilities.

The vertebrates show many morpho-functional variants on a basic theme (Chap. 2). Some of these, such as the pattern of distribution of the genetically distinct chemosensory neurones within the VN epithelium, will be related to the level of complexity of the animal. In some groups, the VNO can be equally complex, whilst the accessory areas of the brain will differ in complexity, as in the advanced reptiles and mammals. Eventually, detailed comparisons of the genomic repertoire of the various accessory systems should reveal the extent of the operational distinctions amongst them. Of particular interest would be the events which account for the suppression of AOS morphogenesis, and those which compensate for its absence.

4.1 PERIPHERAL, CENTRAL AND NON-SENSORY DEVELOPMENT

4.1.1 Peripheral Development

The origin of the nervous tissue which comprises the sensory epithelium of the vomeronasal organ is in the anterior neural crest, from which the anterior neurogenic placode appears at the rostral tip (Fig. 4.1).

The chemosensory stem cells give rise to several types of neuronal and non-neuronal cell lines under the influence of multiple organisers. From a ventro-lateral infolding, the olfactory pit is produced and this invagination soon becomes separated into two areas which will produce the main and accessory olfactory neurones [Figs. 4.2(a)–(d)].

The more ventral zone eventually occupies a position on the lower part of the presumptive nasal septum (Farbman, 1991). These primordia differentiate in the early embryo — at about the 8 mm stage in man. In mammals with short gestation periods, the VN primordium is already established by E12 in rodents, as an invagination which then extends longitudinally (Szabo and Mendoza, 1988). The cigar-like shape of the organ is assembled — as with other tubular structures — through the

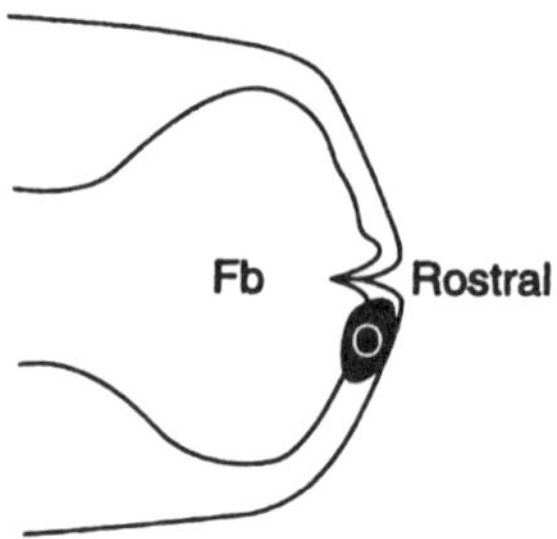

Fig. 4.1 Differentiation of olfactory placode (O). Rostral CNS in vertebrates; Fb = forebrain (from Graham and Begbie, 2000).

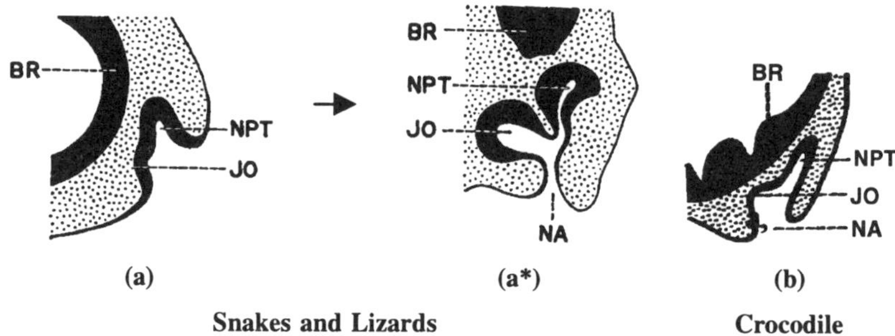

Fig. 4.2 Differentiation from olfactory placode of main and VN primordia: (a) to (b), embryonic stages in reptiles (Squamates and Crocodylia). **(a)**, Early: invagination of placode; **(a*)**, late: separation of primordia for AOS/MOS; and **(b)**, agenesis of presumptive VN(JO) cells in crocodile (NPT, nasal pit) (from Parsons, 1970).

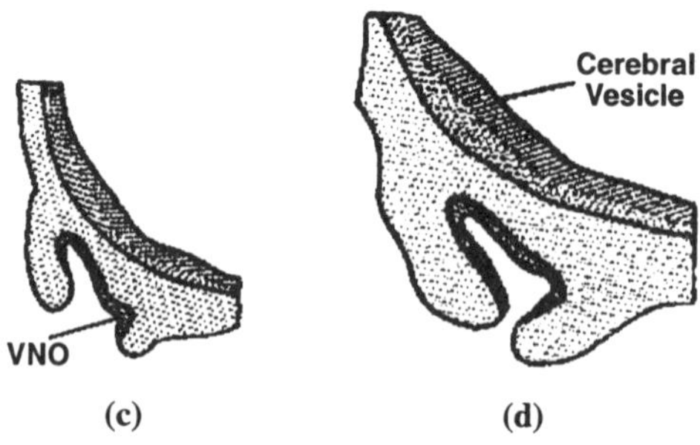

Fig. 4.2 **(c)** to **(d)**, Early and late embryonic stages in mammals, invagination and separation of AOS from MOS (from Farbmann, 1992).

fusion of the lips of a groove, in this case by a "zipper-like" process moving rostrally (Bossy, 1980). On completion of this sequence, the now cylindrical structure remains open only anteriorly as a pore. The entrance to the AOS is always via a non-sensory duct, directly or indirectly related to the naso-palatine duct; in the latter case [Fig. 2.3(b)] it is retained as a separate nasal aperture. The factors which determine the relative position of the VN and NP ducts are the formation of the primary and secondary palates [Fig. 2.3(b)] and their mid-line fusion with the septum. The separation of the two, as seen in rodents, follows from the presence of the "chewing-gap" or diastema between the incisors and pre-molars.

A pseudostratified columnar epithelium now lines the lumen, and undergoes differentiation into the supporting cells and olfactory cells, the nervous and non-nervous components. The other products of the placode are the glial cells and neurocrine (LHRH) cells which locate external to the organ [Sec. 4.4; Fig. 4.6(a)]. The first appearance of maturational changes in the precursor chemosensory neurones of the rat occurs between embryonic days 13 (E13) and 15 (E15) (Garrosa *et al.*, 1986). The early chemosensory cells acquire the distinctive bipolar processes: the dendrite extends towards the lumen, with axonal growth outwards through the *lamina propria* and dorsally along the septum towards the AOB (Graziadei, 1977). This growth phase is accompanied by a steady increase in the mitotic activity of the basal layer (Garrosa *et al.*, 1992). Cell proliferation is continuous up to and beyond birth in most mammalian species. Typically, the maturation of the bipolar cells in the Golden Hamster is complete, judging by their ultrastructure, by post-natal day 10 (Taniguchi *et al.*, 1982). After the initial burst, VN neuronal production is at a peak at about one week after birth in the rat; at this point the neuroepithelium has ten cell layers; but the post-pubertal VNO stabilises with a ~60 μm thick, seven-layered lining. The fully functional organ now contains two-thirds of its receptors as complete bipolar elements — identified by retrograde labelling — this being an increase of some one-third from birth.

VN neurogenesis in Common Marmosets (*Callithrix jacchus*) was detected in the late embryo from about E70 (gestation stage 19) (Phillips, 1976). At birth, these monkeys do show a distinct VNO, but one with little evident neurogenesis (Sec. 4.3 below, Evans, unpubl.). Where the adult AOS is reduced or absent, as occurs in Old-World monkeys, there is a corresponding reduction in early mitotic activity. A low uptake of H^3 occurs in the embryonic VNO of rhesus macaques (Wilson and Hendrickx, 1977). Similarly, the embryonic phase in the monosmic crocodilians shows no differentiation in the ventral region, with active mitosis only in the presumptive main olfactory pit [Fig. 4.2(b)]. The late embryo, in species with a vestigial adult VNO (Chamaeleons), appears similar to that in Fig. 4.2(a); subsequently, neurogenesis does not proceed, and the characteristic ophidian mushroom body fails to develop (Haas, 1947). The chemosensory changes during metamorphosis are not immediately evident in anurans: the VNO of the Reddish frog (*Rana japonica*), remained immature for about one month after hatch. At the end of metamorphosis, prior to the assumption of dual respiration and chemoreception, its ultrastructural development was complete at 60 days after hatch (Taniguchi *et al.*, 1996).

The early VN neurones migrate by various routes and in mammals become organised into two layers, each with distinct functional attributes (Chaps. 2 and 6). Amongst marsupials, the sensory cells of an opossum appear at about one-week post-conception (Jia and Halpern, 1998). Despite their extremely altricial developmental pattern, the bandicoots' dendrites produce sensory processes (Fig. 4.5) on the lumenal border of the VNO at 35 days postnatal, which in the Northern Bandicoot (*Isoodon macrourus*) is about 50% of pouch life (Kratzing, 1986).

In opossums, the migration from the proliferative layer shows a two-day lag between entry of neuroblasts to the lower zone (Go), with

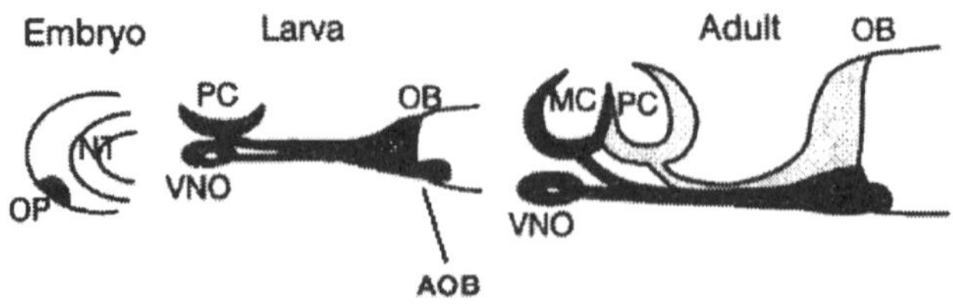

Fig. 4.3 Morphogenesis during metamorphosis in Amphibia: diagramatic sequence of MOS/AOS formation in African Clawed Toad (*Xenopus*). OP, olfactory placode; PC, principal cavity; and MC, mid-cavity (from Meyer and Jadhao, 1996).

a later entry into the upper zone (Gi) (Martinez-Marcos, 2000). As the multilayered epithelium reaches its maximum depth, its boundaries become evident, occupying 50–100% of the lumenal lining (Chap. 2). The position of the major venous sinus will determine the resting appearance of the lumen. Occasional outpockets or bifurcations may be present but rarely, if ever, become directly innervated. Neurogenesis is not completed until the axons pass out dorsally from the capsule and eventually traverse the ethmoid bone's cribriform plate to enter the accessory area of the emerging bulb.

This fan-like spread of fibre bundles is known as the *filia olfactoria*. A small percentage of these fibres exist as a transient and extra-bulbar population (Norgren *et al.*, 1995). The sole function of this sub-set is to act as a guidance or route marker for the presumptive neurocrine cells (c.f. Sec. 4.4). The number of caudally directed axons gradually declines during rat foetal development, leaving only four or five

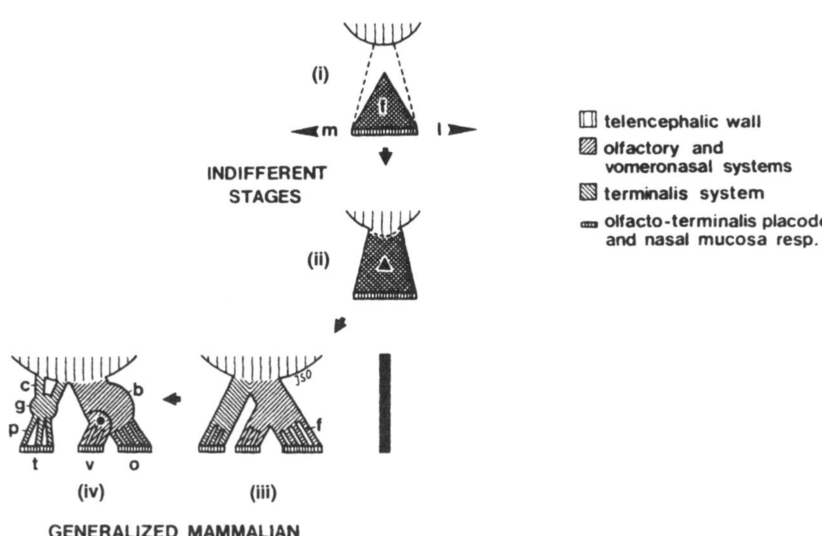

Fig. 4.4(a) Development of the *N. terminalis* in mammals and relationship to *Vomeronasalis* system: (i) f = filia olfactoria; (ii) induction of olfactory bulb; and (iii) to (iv) separation of main/accessory systems (v,o) — from **t**, Nt system; **g**, Nt ganglion + **p/c**, peripheral and **t**, central fibres (from Oelschlaeger, 1989).

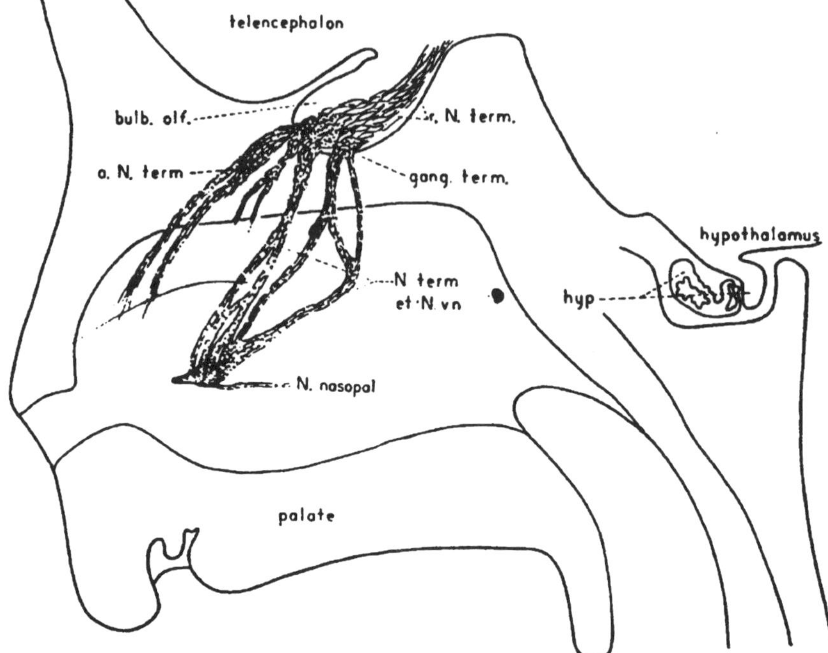

Fig. 4.4(b) *N. terminalis*/Vomeronasal septal innervation in human embryo (45 mm), silver impregnation × 13 (from Pearson, 1941).

present by postnatal day 4 (P4) (Yoshida *et al.*, 1995; Tobet *et al.*, 1995).

In man, the prenatal stages of morphogenesis are still poorly known, particularly from the late foetal through the perinatal period. The fourth branch of the VO nerve has already appeared in the 25 µm embryo (Gulimova, 1996). In the first trimester, neurogenesis is clearly under way by G12 weeks (Ortmann, 1988). Golgi counts reveal a modest population of about 160–180 receptor cells by 18 weeks, and correlates with presence of a moderately specific nuclear marker (NSE), since positive cells are seen within the foetal epithelium up to 12–22 weeks (Boehm *et al.*, 1994; Evans and Grigorieva, 2002, in press). The presence of potentially functional VNORs at this stage, is indicated by the expression of VNR genes in the five-month-old organ (Yukimatsu *et al.*, 2000). The presumed bipolar cells are more likely to be carriers of

Go than Gi receptors (Chap. 6) and their axons also strongly reflect this bias. Additionally, the mid-term stage has emerged as an important phase for non-olfactory contributions, and is also considered below under migration (Sec. 4.4). In its general morphology, the human organ shows a linear growth in size and a logarithmic increase in volume especially between G16 and G30 weeks (Smith *et al.*, 1997).

In cases of abnormal morphogenesis, the clinical tradition of comparison with normal sequences has been applied to the reduction in the human AOS. Cartilage anomalies in patients with cleft lip/palate pathologies are associated with the disruption of foetal VNO formation (Smith *et al.*, 1996). The diminution of the enclosing para-septal cartilage reflects the almost complete absence of the other components, the glandular and vascular tissues, which make up a functional complex. This loss of non-nervous tissue, evident in the embryo, is consistently reported (Humphrey, 1940; Kreutzer and Jatek, 1980; Smith, 1997; Sherwood *et al.*, 1999). The failure of postnatal neurogenesis correlates with the absence of neurone-specific markers (Sec. 4.5 below and Trotier *et al.*, 2000). The loss of the central pathway or its incomplete maturation corresponds to the degree of reduction in the organ. The area of the accessory bulb appears to undergo diminution from the late foetus onward, although mature OR cell markers (OMP) can still be detected by 32 weeks (Chuah, 1987). Careful searches have not revealed any organised bulbar structure in later life (Moran *et al.*, 1994; Meisami *et al.*, 1998). There remains only a possible persistence of some NT, or TriGeminal function, as adjuncts to the VNOR remnant "displaced" to the MOS (Rodriguez *et al.*, 2000). The determinants of this shift in chemosensory function are not understood. A possible model could be that found in a distinction between rat OR/VNOR cells. One of the G-protein variants (G8, = G-subunit 8) is specifically expressed, and persists almost without modification in adult VN neurones. Although potentially active throughout life, this system shows progressive loss of expression in the MOE (Tirindelli and Ryba, 1996).

The gradual suppression of VNR genes and the insertion of pseudogenes (Chap. 6) could account for the differential maturation of the two genomes, eventually leading to the situation where the organ is lost in the adult.

4.1.2 Central Development

Within the accessory bulb, the arrangement and specialisation of the various types of second order neurones is a process which proceeds synchronously with axonal growth. The AOB is still crucially dependent upon its VN input for the production of its mature organisation. The genetic identity of each axon is now thought to largely determine its positional fate within the AOB, as it does for the main bulb (Mombaerts *et al.*, 1999). The precise localisation of OR axons to two of the ~2000 glomeruli exemplifies a highly specific "instructional" growth process (Wang *et al.*, 1998). The expression of OR loci seems to occur during early axonal extension, and necessarily by the time synaptogenesis begins. The multi-causal guidance mechanism is probably a combination of gene expression, axon-associated cell adhesion molecules, and specific growth factors. When the growth tips of the axons enter the bulb's outer layers, they initiate synaptogenesis with dendrites of the newly established second-order neurones, as they extend from the mitral cells to form specific glomeruli. Removal of the Go gene leads to the failure of AOB morphogenesis, since in the absence of this gene, penetration of the posterior bulb is reduced down to the expected level of about 50% (Tanaka, 1999).

Co-localisation of like axons appears to be aided by their "interdependence" during the process of convergence within any one glomerulus. Receptor alleles with related coding sequences (selected by inactivation, Chap. 6) appear to aid the spatially regulated placement (Ebrahimi *et al.*, 2000; Ebrahimi and Chess, 2000). A unique morphogenic feature of the local inter-neuronal arrangements in the glomeruli is suggested as the cellular mechanism for ensuring accurate 3-D placement of "new" axons entering the AOB (Raisman, 1985). Two cell types bracket the presumptive axonal tip in a temporary encapsulation between glial and/or astrocyte processes. Whether these non-neuronal cells contribute the appropriate adhesion molecules preceding the final synaptic contact is unknown. The formation of subsequent "replacement" synapses during regeneration must surely involve the presence of stable cell types. Synaptogenesis, when complete, induces two events: centrally, completion of glomerular internal organisation, and peripherally, activation of the primary neurone. Once the first-order link is established, the primary neurone proceeds with the formation of the free surface of the dendrite, i.e. producing cilia in the MOE and microvilli on the VN dendritic knobs (Menco, 1987; Garrosa *et al.*, 1991).

A mixture of the two receptor cell types differentiates and persists in the only other "secondary" olfactory site in the lower nasal septum (Breipohl *et al.*, 1989; Adams, 1992; Taniguchi *et al.*, 1993). The Septal Organ (of Masera) produces, from about E16 in rodents, axons from this small population of mixed ORs. These reach a demarcated area of the main bulb in the posterior half, mainly restricted to the ventromedial bulbar aspect. The projection pattern appears to be quite distinct from the MOE axons' distribution in the rest of the bulb (Giannetti *et al.*, 1992). Systematic functional studies of this morphologically separate chemoreceptive area are lacking, while comparative surveys have advanced little since its discovery (Rodolfo-Masera, 1943; Kratzing, 1978). Appropriately, a comment on its morphogenesis reinforces the lack of analysis of and interest in Masera's Organ "...a proper chemosensory system with its own time-course of development" (Giannetti *et al.*, 1995b).

Throughout life, all regions of the bulb must have the ability to maintain functional consistency on the arrival of "new" axons designated for main septal or vomeronasal areas. After removal of the target sites for developing axons by bulbectomy, there follows a consequent degeneration of the receptor cells, with subsequent regeneration from the undifferentiated neurones (Barber, 1981; Constanzo, 1991). Acquisition of full capacity is unlikely to be attained, since reinnervation of the bulbar sites is not complete. In the adult, bulbectomy or axotomy has similar effects on the replacement of connections between organ and brain (Roland *et al.*, 1995).

Early transplantation experiments in both lower (frog) and higher (rat) vertebrates with the VN placode grafted into same-age conspecific host, leads to the formation of AOB-like structures. The transplanted axons retain their capacity for organisation of the second-order neurones; expression of their VNOR subset (see above) is the most probable inductive organiser. In non-bulbar areas of the brain, such transplants induce glomeruli in atypical (ectopic) positions (Graziadei, 1990).

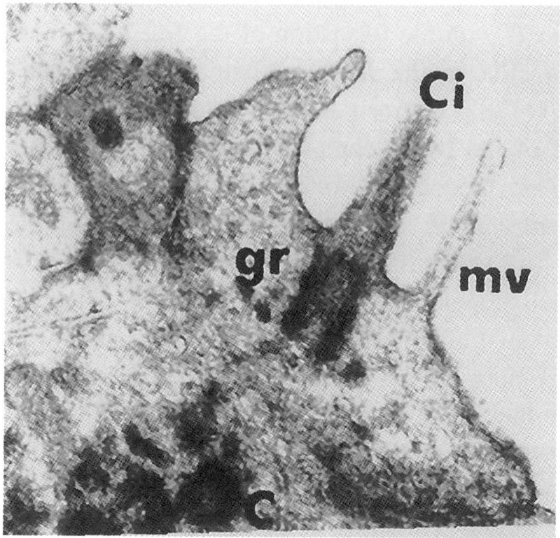

Fig. 4.5 Receptor cell morphogenesis: dual projections, presence of cilia and microvilli on an early VN cell. Thin section of Northern Bandicoot (*Isoodon macrourus*) — 35 days postnatal (= 50% pouch life); C, centrioles; Ci, cilium; gr, dark granules; mv, microvilli; TEM; Scale bar: 1.0 μm (from Kratzing, 1986).

The output axons of the bulb emerge from its caudolateral aspect to make up the lateral olfactory tract leading to the AOS tertiary structures, most immediately the anterior olfactory nucleus.

4.1.3 Non-Sensory Development

The VN cartilage, the VN glands and the vasomotor network are all functionally necessary components without whose various contributions the receptor cells would be unlikely to function. These ancillary constituents show a parallel developmental sequence. The cartilage may become partly ossified and incompletely surrounds the organ, the enclosure lessening towards the posterior end. In the mature organ, maximum enclosure is usually found along the sensory section.

In the posterior glands of the nasal septum and in the vomeronasal glands an odorant-binding protein (OBP-II), as expected, increases during postnatal stages when the need for carrier compounds (Chap. 3) rises (Ohno *et al.*, 1986). The sero-mucous secretions are required to form the protective peri-microvillous layer at the free surface. Within the mucomicrovillar complex (MMC) and within VN glands of neonate, as well as adult rats, detoxification of xenobiotics (and possibly the clearance of pheromones) is attributed to the components of the γ-glutamyl cycle (Krishna *et al.*, 1992). These perireceptor operations protect the local environment, and may prevent premature activation of the AOS on exposure to sexual chemosignals.

The vasculature is established by the 18th day of gestation in rats, and comes from the arterial supply as the anterior cerebral vessel, eventually entering the basal lamina via septal tributaries of the olfactory artery (Szabo, 1988). Unlike the MOE, the organ's capillaries penetrate in loops into the neuroepithelium. Blood from the vomeronasal complex arrives for collection in the vomeronasal vein, as described earlier [Fig. 2.11(a)]. The establishment of the highly vascular columnar complexes seen in the ophidian organ has not been correlated with functional development (c.f. Wang and Halpern, 1980; Holtzman and Halpern, 1990).

4.2 CONTINUOUS AND DELAYED NEUROGENESIS

New neurones are produced as replacement elements for the senescent cells, which are presumed to wear out during a lifetime's intake of deleterious molecules. This unique provision of neuronal turnover maintains the optimum density of the mature receptor population throughout life (Barber and Raisman, 1978; Wilson and Raisman, 1980). The rate of maturation of the somewhat protected VN olfactory neurones is generally slower than those of the more exposed primary tissue (Moulton, 1970; Celebi *et al.*, 1970). The dynamic state of the neuroepithelium means that it contains several stages of immature (non-functional) receptor cells present at all times, and at various levels (Graziadei, 1990). A reserve population of precursor neurones is thus available for response to appropriate activational signals; the local density of mature neurones is one regulatory factor which feeds back to the progenitor cells. Both inhibitory and activational signals are considered to control the rate of neurogenesis (Calof *et al.*, 1998). Intra-epithelial repair-responses to cell damage show that after axon cuts, all cells of the MOE become sensitive to the NGF signal by receptor induction activity (Miwa *et al.*, 1993). The regeneration ability of VN neurones in response to damage is similar to that of the MOE, as is its mitotic capability (Barber and Raisman, 1978a and b). The normal cycle time for complete replacement, i.e. up to AOB synaptogenesis, is not known; senescent decline may exist, or a consistent rate of re-growth may be maintained. The effects of VN-axonal cuts are quite variable in that only 30% of new (labelled) fibres reached the AOB (Ichikawa, 1999). The process lasted up to eight months, and was evident in only a third of the operated rats. The normal replacement cycle presumably maintains a core population which covers all receptor categories uniformly and prevents "genetic gaps" appearing in vomeronasal responsiveness.

At least two pools of presumptive neurones are proposed (Martinez-Marcos *et al.*, 2000). One which initiates the first population of receptor cells comes from the strip of basal mitotic cells along the lamina. The other group comes from a more lateral cluster at the boundary with the non-sensory epithelium. The reserve pool(s) have been found to migrate to a central position, where they enter a resting phase until activated. By analogy with the main receptor cells, the rate of VN neuronal turnover could presumably be slowed considerably by "cleaning up" the organ's (fluid) intake. In the rat MOS, filtered air slows, or even prevents, the replacement of receptors (Hinds *et al.*, 1984). However, the unilateral removal of most volatile stimulation, by blockage of one nostril, is without effect on the developing (ipsilateral) AOB (Brunjes and Kishore, 1998). No experimental data has been produced to determine whether completely preventing stimulus access alters VNE and then AOB morphogenesis and regeneration.

A cognate property of this continuous production throughout life is the repair of damaged parts of the neurone: axonal and or dendritic re-growth readily occurs, an ability retained even by higher primates (Graziadei *et al.*, 1980). Such a regenerational capacity is probably due to the peripheral site of the cell body, since those within the CNS mostly lack this property.

4.2.1 Delayed Neurogenesis

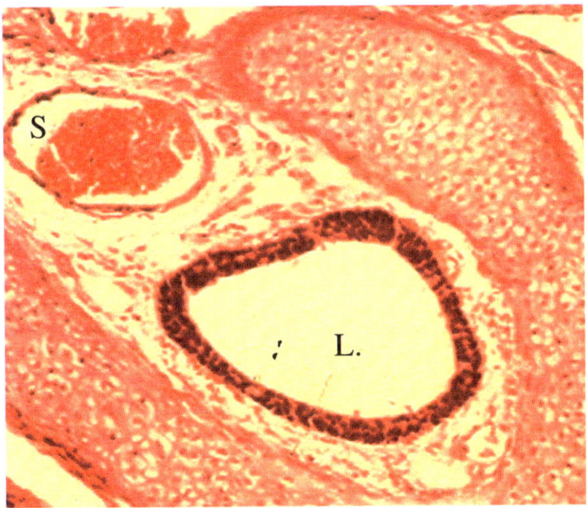

Pl. 4A Neonate Prosimian: undifferentiated VN epithelium in 11-day-old Slow Loris (*Nycticebus coucang*) (from Evans and Schilling, 1995). L. = lumen, s = sinus.

A few primates start life with a VNO whose neonatal epithelium is almost completely unstimulated, with little or no apparent build up of receptor cell density. Slow neurogenesis occurs in some species of Prosimians and in the Marmoset/Tamarin group. The entire organ of the neonate has a duct-like appearance, with up to three rows of undifferentiated cells along its length. Serial sections taken at full-term in a Slow Loris (Pl. 4A above) show an epithelial lining without a dendritic layer, but surrounding a well-formed lumen (Evans and Schilling, 1995). Vascular sinuses are conspicuous within the cartilage, although the complex lacks a glandular presence. In the Marmoset/Tamarin monkeys, the neonates of several species show a restricted neurone-specific lectin binding in comparison with the VNO of adult conspecifics (Taniguchi *et al.*, 1992; Evans and Grigorieva, 1995). A fully competent adult state has been confirmed at LM level in about 95% of Prosimians; indeed the adult Ring-tailed lemur (*L. catta*) grows a VNO with a complete neural lining, uninterrupted by a non-sensory zone (Evans, 1991 and Pl. 2.1B). Other structural surveys of Callitrichidae also suggest a functional status for the adult organ, as do the effects of vomeronasalectomy (Hunter, 1984; Barrett *et al.*, 1993a and b; Mendoza *et al.*, 1994).

4.3 STIMULUS ACCESS

Prenatal entry of odourants from the amniotic fluid to the VN lumen and then into the peri-receptor mucus sheet is presumed to be facilitated by the circulation of this all-pervasive bathing medium. Swallowing movements may well assist the transport of whatever substances are introduced by trans-placental movement from the maternal environment. Several studies suggest that the olfactory systems develop an early capacity for response to stimulation, even in altricial species. Prenatal exposure of foetal rats to odourants can be shown to affect aversions or preferences during suckling, although the stimuli may be processed through either the MOS and or AOS (Pedersen *et al.*, 1983; Smotherman, 1992).

Limitations on stimulus access *in utero* may present problems for interpretation of results in such experiments. Prenatally, the rat's VN ducts are already patent (Coppola and Millar, 1994). The AOS cannot then be ruled out as a probable candidate pathway for effective prenatal stimulation, especially if neurogenesis is underway, if not complete. By contrast, the foetal domestic mouse has a VNO in which access is apparently sealed off, with sloughed cells filling the lumen of its duct. The entrance only gradually becomes patent just after birth, but is still not fully accessible, being "immature" until day 25 (Coppola *et al.*, 1993). The chance of prenatal stimulation occurring in mice is not high, since access tests with microspheres (0.95 μm diameter) fail to show uptake and hence support this contention. Whether the transient blockage of the duct system(s) is a consistent feature of AOS development cannot be generalised until further comparative evidence emerges. Supportive findings in an early ungulate embryo are that in the domestic pig "…up to 57 mm, T.L… ducts and furrows of the vomeronasal complex were blocked by fused epitheliums" (Wöhrmann-Repenning, 1994). Again at birth, the pig VNO is fully mature by TEM criteria (Kratzing, 1976), suggesting that a transient blockade is

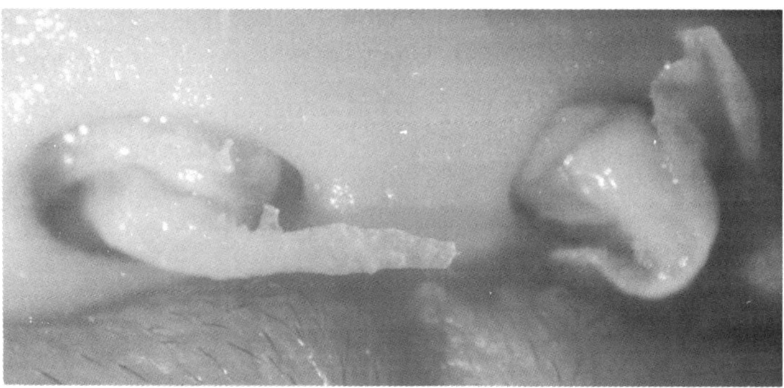

Pl. 4B Prenatal occlusion of nostrils in man: nasal "plug" formation during foetal life (c.f. Schaeffer, Fig. b, 1910); mid-term ~ 16/17 weeks/HC, 96 μm (Museum No. 62/1048 U.B.).

emerging as a common feature of morphogenesis. Even if incomplete, such restrictions of fluid access to the lumen could well limit early responsiveness.

In man, an overlooked feature is the occurrence of mucoid-like plugs in the foetal nostrils (Schaeffer, 1910). The presence of this blockage can be confirmed by endoscopic inspection *in utero*; these plugs seem likely to affect free amniotic flow, since they appear to be reinforced by a folded membranous gathering at the nasal vestibule (Pl. 4B). A degree of restriction of fluid access to the VN aperture, which is immediately caudal to the nostril aperture, and is patent in foetal life, may be a protective feature (Jordan, 1972). The timing of the dissolution of these "sealant" devices prior to parturition is regrettably not known.

Later in intra-uterine life, the human infant is susceptible to early chemical prompting, but again the affector route is not known with certainty. Neonatal discrimination in favour of familiar (maternal) amniotic fluid is demonstrable, suggesting that the foetus already has active chemosensory capacities (Schaal, 1998). Smell and taste are operative in the near full-term foetus since it shows detection of about 120 mg/day maternal intake of anethole (as anise condiments) within a few days before parturition; this exposure induced subsequent preferential responses by babies to anethole (Schaal *et al.*, 2000). The human neonate is not likely to have its organ as a fully functioning chemosensor, but may still utilise the remnant VNORs within the VNE and the MOE at this period (Rodriguez *et al.*, 2000; Yukimatsu *et al.*, 2000).

The apparent variability of prenatal odourant access and difficulties inherent in assigning any specificity of later response to a defined early stimulus can be partly circumvented by inferences from the immediate post-parturition phase. In altricial species, the immobile and sensorial reduced neonates may not be equipped to employ contact or close-range olfaction. For key responses such as nipple-location, rabbit pups may use volatiles only; indeed the orientation to the doe for episodic suckling does not seem to be affected by VN-X procedures (Hudson and Distel, 1986; Keil and von Stralendorff, 1990). The role of the organ in altricial neonates is only one part of a complex interplay between offspring and mother involving close-range chemocommunication (Brouette-Lahlou *et al.*, 1992; Vernet-Maury *et al.*, 1993). In relatively precocial neonates, some familiarity with maternal (e.g. skin) odours could enhance attachment, initially established by previous exposure to background stimulation from the contents of amniotic fluid (Schaal, 1995). Similarly, maternal responsiveness to lamb odours is also assisted by the presence of amniotic stimuli, although the degree of dependence on the AOS and/or MOS is not fully established (Levy *et al.*, 1987; Booth *et al.*, 2000). Early attachment to the milk source rapidly becomes individually specific, even in apparently microsmatic humans. However, responses to other social odours can be VN-independent in ewes, relying largely on the MOS (Cohen-Tannoudji *et al.*, 1989). Although early social contacts frequently involve licking and/or nuzzling between mother and offspring, these behaviours again need not necessarily activate AOS input (Levy *et al.*, 1995; Roland *et al.*, 1995).

4.4 NEUROCRINE CELL MIGRATION

The direct link between the nose and the reproductive system has been called “a non-cognitive bridge” (Stoddart, 1990). A part of VN projection is non-olfactory and reaches the limbic system via a direct and monosynaptic relay (Larriva-Sahd *et al.*, 1993 and 1994; Matsumoto *et al.*, 1994). In addition to the production of the peripheral olfactory cells (Sec. 4.1 above), the placode also gives rise to a group of important

CNS neuroendocrine cells. The gonadotrophin-releasing hormone cells of the mature hypothalamic regulatory nucleus and the medial preoptic area have their origin in the epithelium of the medial placode, and are found amongst its early derivatives (Witkin, 1987; Schwanzel-Fukada and Pfaff, 1989). The origin of GnRH-secreting neurones in a relatively remote site could well reflect the basic pattern of morphogenesis and initial organisation of the early vertebrate nasal-pituitary-gonadal axis (Knouff, 1935; Klein and Graziadei, 1983). As with non-secretory neurones, there is a modelling phase during growth, in that embryonic over-production is followed by a postnatal reduction of about 55% in the nasal, plus the CNS population of GnRH-staining cells (Wu *et al.*, 1997).

The ontogenesis of the AOS then, is closely bound up with the formation of its principal connection site within the mature brain. The sequence of events in mammals is revealed as a process which involves: (1) early specialisation of presumptive GnRH cells; (2) their attachment to and movement along specific (and transient) axonal bundles of the VN and *N. terminalis* tracts, and (3) coalescence of the neurocrine cells in the hypothalamus, where they complete differentiation as multi-axonal neurocrine cells.

This rostral "migratory stream" is thus guided through the cribriform plate and forebrain to allow formation of the gonadotrophin-regulating nucleus. Part of the guidance mechanism involves the expression of N-CAM, one of the membrane-adhesion molecules, such that the fascicles of the nerve(s) are spaced to allow passage of the GnRH cell chains (Schwanzel-Fukuda *et al.*, 1994; Schwanzel-Fukuda and Crossin, 1996). The delivery route to the (rat) forebrain is made up of a sub-population of extra-bulbar dispersed fascicles which regress soon after birth (Yoshida *et al.*, 1995) [Fig. 4.6(a)].

The human AOS still follows the mammalian pattern, as GnRH-positive cells appear in the VNO at the same stage — 21–25 mm C/R, between six to eight weeks gestation — as the emergence of VN nerve branches (Boehm *et al.*, 1994; Kjaier and Hansen, 1996a; Gulimova, 1996). As the GnRH cells mature, they lose an embryonic membrane glycoprotein which appears in the placode, but persists as a marker in adult olfactory neurones (Okabe *et al.*, 1996). The inductive properties

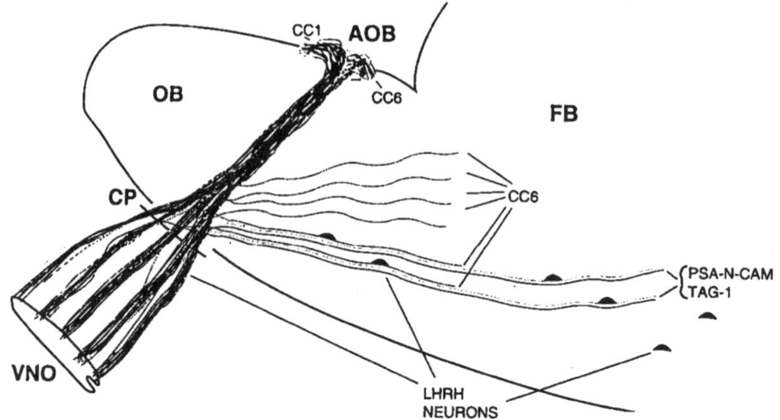

Fig. 4.6(a) Migration of LHRH neurocrine cells: prenatal transportation along the track of extra-bulbar VN axons (caudal branch). CB, cribriform plate; FB, forebrain; cell types, TAG-1, transient axonal surface glycoprotein; and N-CAM, neural cell adhesion molecule (from Yoshida *et al.*, 1995).

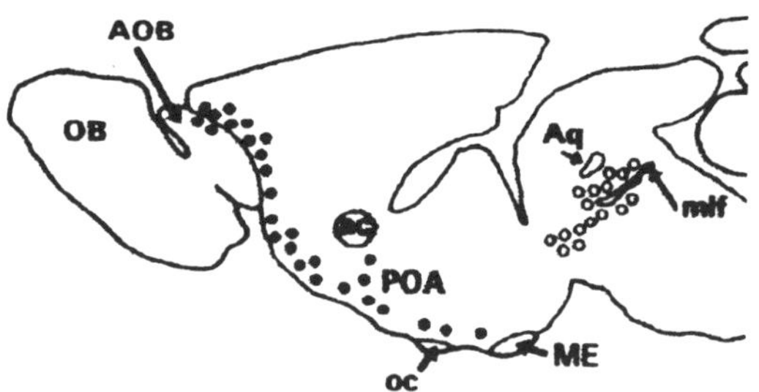

Fig. 4.6(b) GnRH +ve neurones (relative density) in the brain: dot = mammalian, circle = avian, forms. Parasagittal view of distribution in Musk Shrew (*Suncus murinus*) — from bulb to median eminence (ME), optic X (OC), cerebral aqueduct (Aq), and preoptic area (POA) (from Schwanzel-Fukuda and Pfaff, 1994).

of the early hypothalamus provide further support for the exertion of a degree of central to peripheral influence. Explants of cells from its medio-basal nucleus were the most effective in inducing the *in vitro* migration of GnRH neurones from the rat VNO primordium (Daikoku *et al.*, 1993).

Confirmation of a link between the establishment of the olfactory and gonadal systems comes from a rare pathological X-linked state in the human male (Legouis *et al.*, 1994). Adult anosmia is associated with sterility: Kallmann's syndrome — attributed to the morphogenic failure of the bulb and a pre-migrational arrest of the placodal GnRH cells (Quinton *et al.*, 1997). The guidance axons lack any target, hence cannot achieve synaptogenesis and in consequence do not survive to assist with neurocrine cell transport. The loss of GnRH hypothalamic nuclei is a result of agenesis in the bulb, since the anterior pituitary itself remains competent, but is deprived of its regulatory hormone and thus fails to direct testicular maturation. Alternatively, partial failure of adhesion molecule expression could explain some cases of hyposmia with an AOS component.

4.5 FUNCTIONAL MATURATION

The pituitary, as well as the hypothalamic hormones, also contribute to VN development. A transient prolactin receptor (PRLR) is expressed in the late foetal rat. At E18, there is positive staining for this binding protein along the lumenal border; the reaction is restricted to the medial (sensory) zone [Freemark *et al.*, Fig. 6(d), 1996]. These sites possibly function in the detection of endogenous lactogenic ligands such as PL-I and PL-II. The VNORs also occupy microvillous sites in this area, but the precise developmental role of PRLR in the early AOS is unknown. The modulatory influence of prolactin is well established after puberty in mammals as a reproductive determinant (Chap. 5). In a more central role, it acts on the EOG recorded from the accessory area of the bulb in newts (Toyoda, 2000).

A degree of sexual dimorphism (Figs. 5.9 and 5.10) has been shown in primary to central units of the rat AOS (Madeira and Lieberman, 1995; Guillamon, 1997). It becomes morphologically distinct (male > female), with more and larger neurones produced throughout the system. Administration or removal of androgens affects the degree of neuronal development within the VN epithelium, the AOB and central nuclei such as MeA and BSTe of the amygdala (Simerly, 1990). As expected, there are numerous sites where gonadal steroid receptors are expressed in the central nuclei (Guillamon and Segovia, 1997). As will be seen (Chap. 5), sexually distinct signals are able to elicit sex-typical responses specifically from the AOS. In the rodents, the AOS is strongly implicated in sex-typical responses, but it is not present in a dimorphic condition in all other mammalian groups.

Histo- and immuno-chemical changes identify events which underlie intra- and inter-cellular differentiation. Cell-cell interactions are very likely to be influenced by the particular properties of neuronal membranes (Sharon, 1989; Astic, 1989). In addition, the extracellular matrix can be characterised by the presence of carbohydrate-containing glycoproteins. These non-enzymic proteins selectively and reversibly bind certain mono- and oligo-saccharides; sugar-specific binding is visualised through labelled plant lectins. In a few species, UEA-1 (gorse: *Ulex europeus*) and SBA (soy bean: *Glycine max*) recognise only olfactory cell membranes (Key and Giorgi, 1986; Barber, 1989). Binding by UEA to the α-L-fucose group identifies an olfactory membrane glycoform, one of the neural cell-adhesion molecules (N-CAM) with importance for axonal growth (Pestean *et al.*, 1995). By contrast, in a broad spectrum reaction (to RCA-1), localisation occurred in non-sensory tissues: the secretory granules of sustentacular cells, in acinar cells of nasal glands and at the mucocilliary surface. The common binding component in this case was identified as the glycoprotein Olfactomedin (Snyder *et al.*, 1991). Specific glycoconjugates identify the secretory granules of the VN glands, recognised by α-D-galactose and β-N-acetyl-D-galactosamine residues, as with the VN neurones themselves (Takami *et al.*, 1994).

Overall, a wide range and degree of affinities have been found, often showing inter-specific variability. Nevertheless, lectin-binding can be used to identify changes with age in the binding reactions of primary and secondary AOS neurones. The two human systems exemplify such changes, since between G8 and G12 weeks the UEA-1 reaction in the MOE appears only from the 11th week, whereas in the VNE it is detectable at week 14, continuing to at least week 22 (Gheri-Bryk *et al.*, 1992; Evans and Grigorieva, 2002, in press). Ontogenetic changes in neuronal membrane characteristics occur in the mouse opossum, in laboratory rodents and in some callitrichid monkeys (Franceschini *et al.*, 1994; Evans and Grigorieva, 1995; Shapiro and Halpern, 1995).

This process seems to reflect gradual increases in the intensity and density of labelled receptor cell bodies and their axons, followed by regional bulbar staining as synaptogenesis proceeds. However, a more extensive range of lectins needs to be examined before the implications of species differences in the membrane glycoproteins can be satisfactorily interpreted (Salazar and Quinteiro, 1998).

Neuronal markers expressed only in mature cells (e.g. OMP), help in the estimation of the relative status of peripheral or central cells, and of transient versus completed stages (Rogers, 1987; Baker, 1988; Johnson, 1993; Shapiro, 1997). The identity or otherwise of glycoproteins with receptor sites will be required before changes in structural characteristics can be related to functional morphogenesis.

As with most cells exposed to environmental agents, the olfactory cells require protective degradation enzymes to remove potentially hazardous molecules. A postnatal increase in one of the dismutases which eliminate free radicals and other xenobiotics, again correlates to exposure of the nose to non-odorous chemicals (Kulkarni-Narla *et al.*, 1997). The activation process begins in the exposed dendritic regions; the microvilli of neonatal rats already possess compounds which degrade xenobiotics (Krishna *et al.*, 1992). These authors speculate that a similar role is responsible for their presence here in acting to remove redundant or unbound chemosignals. Overall, the enclosure of receptor cells within the VN does not seem to give them any significant added protection, despite their limited direct exposure to the atmosphere. Only with a highly volatile gas such as ammonia (NH_3), is there differential penetration; the organ's neuroepithelium is subject to a lesser level of destructive effects than in the MOE of rabbits (Gaafar *et al.*, 1998).

4.6 GENERAL

Species differences indicate that the AOS is not invariably operational prenatally, even though most peripheral and central neural units are in place and available for activation. Variability in the timing of maturation of the Organ-to-AOB linkage could well provide the necessary flexibility of response consistently associated with higher mammalian, and especially primate, neural systems. The onset of effective accessory chemoreception varies from perinatal to almost peri-pubertal. Hence, full developmental sequences are required before reliable interpretations of later preferential responses can be made. The specific attribution of VNO response to the pattern of chemicals present in amniotic fluid for example, is clearly desirable. Both olfactory systems seem to have the potential for a role in sexual differentiation of the brain, but information of the extent and timing of any involvement is tantalisingly scarce.

Postnatal modelling by means of specific alterations to bulbar circuits during growth is undoubtedly a key mechanism in adaptive responses. The establishment of AOS-dependent responses in the adult relies on the ability of the sensory processing circuits to adjust to changes in chemosignal input at a local level. A well-established analysis of such a model incorporating an effective and brief critical period, is the odour-imprinting events occurring during the Bruce effect (Keverne and Brennan, 1996). The AOB is seen here as a flexible system which alters its inhibitory processes to accommodate ongoing social situations (Chap. 5).

The AOS may also make an important contribution to that portion of conspecific identity that relates to MHC-typing. Should it be fully established that some inter-individual responses are AOS influenced — wholly or partly — then this type of chemoreception assumes long-term significance. The accessory system's phylogenetic persistence could rest on its role in providing a means of inbreeding avoidance and/or mate-selection for heterozygosity amongst related individuals. However, this assumption raises questions about the monosmic groups, those lacking an adult AOS: does the MOS take over and provide a similar function and if not, are there substitutes? Several species without an AOS retain the NT network, presumably to maintain the delivery of releasing-factor cells to the hypothalamus. Other functions, such as local actions of LHRH, await further research.

The developmental changes seen in the immediate postnatal period in altricial rodents and especially in the early stages of marsupials, are an expected outcome of their shortened gestational period, early parturition and consequential dependent status. Regrettably, the relative contribution of the main and accessory chemosensory route(s) cannot be fully assessed. The lesser importance of the AOS (by some tests) in ewes suggests that assumptions about parent offspring signalling, necessarily involving one or other system, are premature, particularly since later findings have emphasised its role (Cohen-Tannoudji, 1989; Booth and Katz, 2000). An alternative strategy is to extend the, at present limited, studies on the AOS of altricial and precocial species. Comparisons of other natural variants, such as induced versus spontaneous ovulating species, could illuminate the activation or inhibition of the AOS and its interaction with hormonal status. The avoidance of inappropriately timed endocrine secretion is likely to be the principal contribution of VN input to reproductive organisation and activation.

In support of this contention, the carrier protein Aphrodisin makes an early appearance in vaginal secretions. In pre-pubertal hamsters, it thus indicates chemosensory preparation for the onset of female maturity (Magert, 1999). The proven ability of the AOS to modulate the CNS-pituitary-gonadal axis by advancing or retarding endocrine activity (Chap. 5), underlines its role as primarily the chemosensor of the reproductive system. The adaptive consequence of responses, which allows an avoidance of premature breeding, or of a postponement of puberty, would seem to be advantageous.

5 *PHYSIOLOGY*

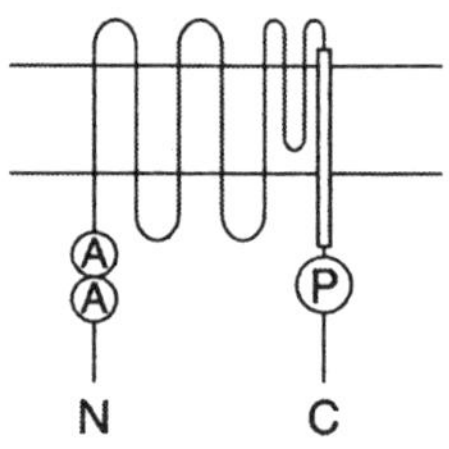

Short TRPC-2

VN Ion-Channel (from Harteneck *et al.*, 2000).

"Quels seront les effets qui se manifesteront lorsqu'on détruira l'organe?"

Louis Jacobson (1812)

5.1 ACTIVATION OF THE AOS

As with any nervous system, the way in which the AOS functions is regulated by developmental and heritable factors (Chaps. 4 and 6), whose interplay determines its operation under any given conditions. Some of its processes can vary with the prior experience of the individual; others are almost invariant and appear as stereotypical responses with little experiential influence. They range from the transient guidance role in GnRH-cell migration, to the alteration in the timing of puberty and the diminution of fertility in adults. Differences in the relative importance of AOS or MOS *can* be considerable, and a complete separation of their contributions is often not possible. The degree of interaction between the systems probably also varies with the type of input presented for processing (Meredith, 1998). The central feature of AOS function is its ability to regulate the activity of the gonads; this it does through the neural pathway influencing hypothalamic control over gonadotrophin secretion (Bellringer *et al.*, 1980; Lepri and Wysocki, 1988; Martinez-Marcos *et al.*, 1999). An unresolved question is the extent of any contribution in animals where the AOS does not reach functional maturity (Chap. 1). It needs to be established whether and by what mechanisms the groups in which adults lack an AOS develop alternative sensory capabilities.

The source of excitability in the organ's endothelium has been examined in the Garter snake by comparing the depolarisation of its VNE after exposure to a neutral (control) compound (n-amyl acetate/ glutamate), and to the skin secretion elicited from earthworms. The prey-derived mixture produces a differentially greater excitation which is lost as the neurones degenerate following damage to their axons (Taniguchi

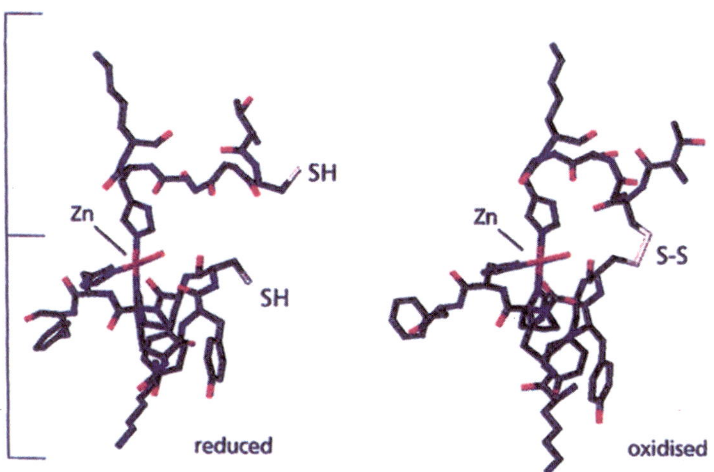

Pl. 5.1 Receptor activation: a model proposed for the induction of OR conformational changes following odourant attachment to binding site on ciliary membrane (from Turin, 1996).

et al., 1998). Since this procedure leaves the supporting cells intact, the electrical activity in the organ's lining stems from the neurosensory component. On activation of the olfactory cell membrane, it initiates a generator potential across the neuroepithelium; the current flow (Fig. 5.1). within the receptor sheet produces, by summation, the electrolfactogram (EOG).

The binding of an odourant to the 7-trans-membrane receptors (GPCR) found on the olfactory cilia and vomeronasal microvilli is considered in Chap. 6. The model visualised in Pl. 5.1 represents the

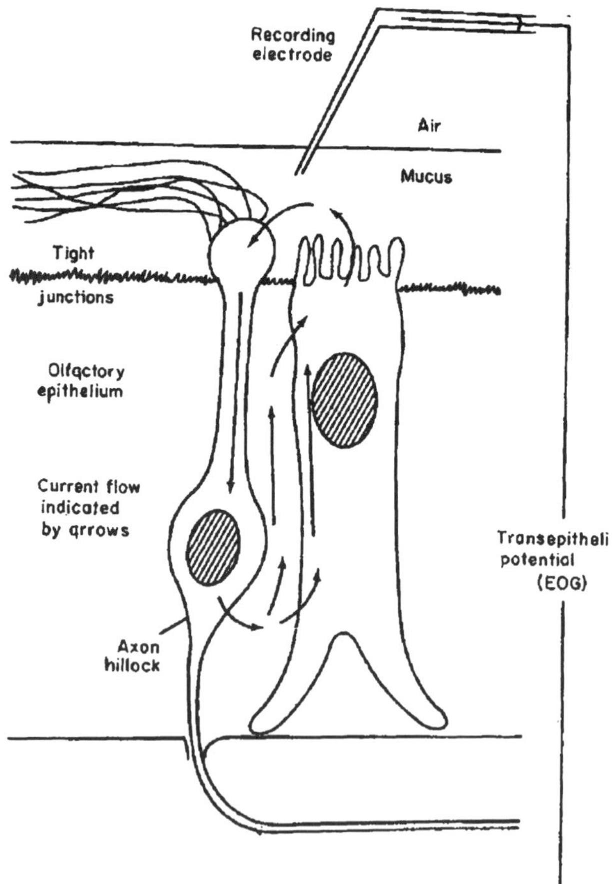

Fig. 5.1 Generation of the EOG: current flow across the olfactory neuroepithelium (from Gesteland, 1971).

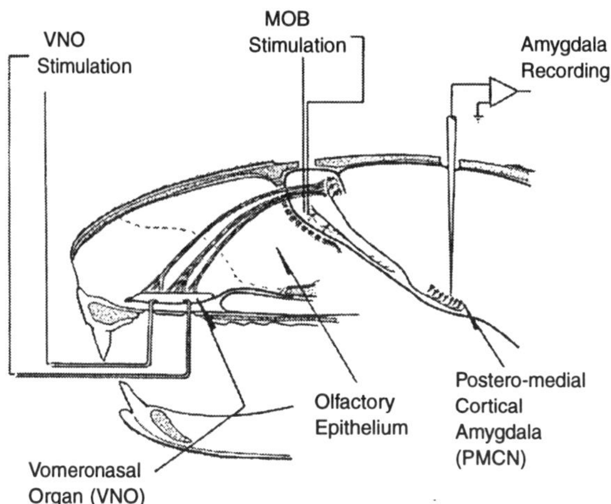

Fig. 5.2 Recording and/or stimulation in Peripheral and Central AOS: electrode/injection sites at primary (VNO), secondary (AOB) and tertiary (PMCN) levels (from Licht and Meredith, 1987).

binding event as one which acts to permit electron flow (Turin, 1996). Activation of the transduction process is conceived as related to the "electron tunnelling spectra of odorant molecules", and olfaction as thus akin to light and sound detection. The Turin model may also apply to the transduction process for vomodours. While the extracellular processes such as lipocalin-mediated delivery of ligands (Chap. 3) may be similar, the intra-cellular events (see below) do differ. Within the organ, a vomeronasal-EOG is produced by the basic changes in membrane conductance (Breer, 1994; Liman and Corey, 1996). The VN neurones have a resting potential of about −50 mV: ranging from −35 mV down to −75 mV, as measured by patch-clamp recording (Moss *et al.*, 1998). When active urinary signal compounds such as dehydro-exo-brevicomin (DHB) are applied to dissociated cells, they produced changes consistent with ion-channel closure (Moss *et al.*, 1998). The proposed mechanism is that DHB and related compounds act to reduce membrane conductance and thus suppress the production of action potentials (a.p.) in the VNO. Depolarisation of isolated rat VN neurones activates a fast transient

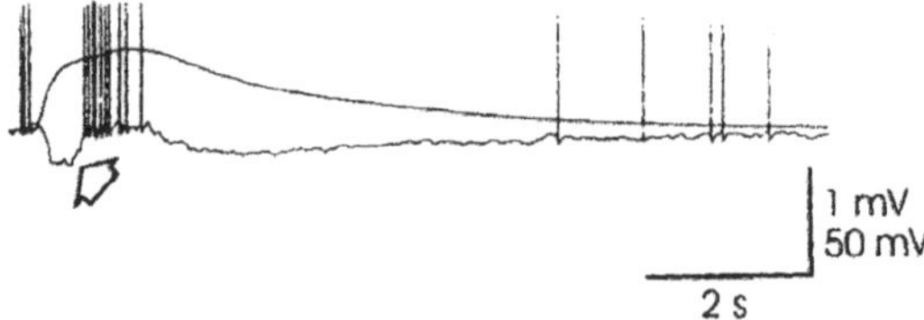

Fig. 5.3(a) Post-synaptic responses of Salamander bulbar neurones to a single odour pulse (arrow), at nostril (from Kauer, 1991).

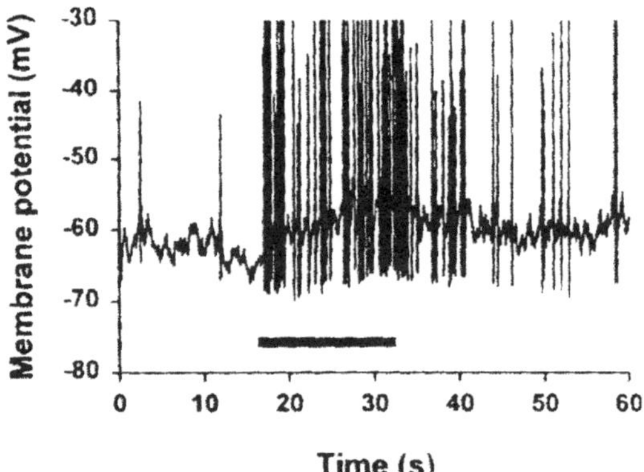

Fig. 5.3(b) Female rat, isolated VN neuroepithelial cell: response to 1:20 dilution of male rat urine, spike traces truncated at −30 mV (from Trotier *et al.*, 1998).

sodium inward current and a sustained potassium outward current (Trotier *et al.*, 1998). The a.p. burst induced by (diluted) male urine [Fig. 5.3(b)] follows the time course of the stimulus when applied to female VNOR cells (Holy *et al.*, 2000). Selective tuning of receptor sites to sex-specific signals is strongly suggested by these findings (c.f. Fig. 5.11).

The intracellular processes which precede membrane activation appear to differ from those of MOE neurones, in that cyclic nucleotide gating may not occur. The transduction process which induces current flow in snake VN neurones, utilises as a putative second-messenger the modulator compound inositol triphosphate — Ins. (1,4,5) P3 = IP3 (Liu *et al.*, 1999; Taniguichi *et al.*, 2000). The proposed channel component associated with the microvillous membrane is one of the transient receptor potential family (TRPC-2: Heading Fig., pp. 94), the β-splice version being VNE-specific in mice (Liman *et al.*, 1999). The enzymic step most closely coupled with this channel type is activation of the PLC cascade (phospholipase-C) (Harteneck *et al.*, 2000). Support for PLC involvement comes from the prevention of VNE responsiveness to

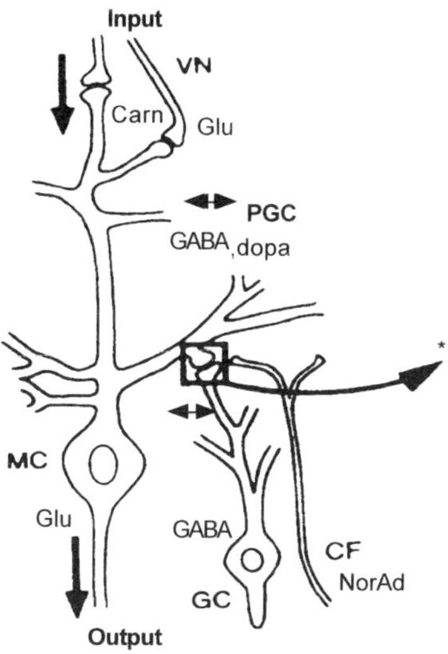

Fig. 5.4 Location of major neurotransmitters in AOB. VN Input to mitral cells (MC) modified and processed by interaction with Periglomerular (PGC) + Granule (GC) cells; ⇔ = reciprocal synapses; efferent/centrifugal (CF) input from ventral sympathetic fibers (c.f. Fig. 2.20), Output via lateral olfactory tract. *Box details in Figs. 5.12(a) and (b) (after Kaba *et al.*, 1990; Shepherd, 1997).

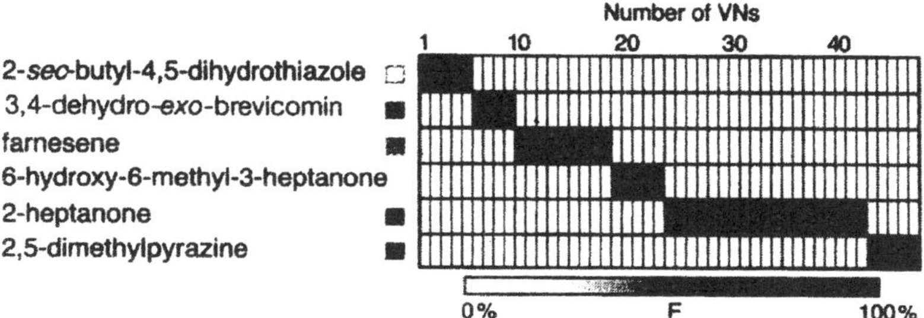

Fig. 5.5(a) Segregation of individual VN receptors: responsiveness to six chemosignals present in mouse urine, non-overlapping pattern demonstrating neuronal selectivity for specific ligands (from Leinders-Zufall *et al.*, 2000).

urinary stimuli by specific inhibitors of PLC (Imamura *et al.*, 1997; Holy *et al.*, 2000). A comprehensive treatment of the details of transduction mechanism in the neurones of the MOS is given by Schild and Restrepo (1998).

At the first transmission site, the AOB's mitral/granule cells utilise Nor-adrenalin (NorAd) as a principal neuromodulatory element, along with γ-amino butyric acid (GABA) and glutamate (Glu). Blockade of NorAd's action by an inhibitor (phentolamine) prevents the recognition of strain-specific male urinary signals in the bulb (Fig. 5.13 below). The bulb also contains peptides such as substance-P (SP) and Met-enkephalin (ENK-8); in several cells these are co-localised (present in the same neurone). Some 15–20% of granule cells contained both compounds (Gouda *et al.*, 1990). Enkephalin is present in the MOB but without co-localisation; the distribution of ENK could represent a functional distinction between the bulbar areas. Variations in cholinergic and non-cholinergic input may underlie the AOB's glomerular handling of vomerolfactory information in primitive mammals such as European hedgehogs (*Erinaceus europaeus*). Enzymatic staining revealed a dual fibre pattern for choline acetyltransferase and acetylcholinesterase (Crespo *et al.*, 1999). In general, the insectivore innervation was similar to the rodent AOB but displayed fibre bundles of both types.

The organ also contains neuroactive compounds as constituents of the vasomotor and neuro-glandular tissues (Zancanaro *et al.*, 1997). These include the amine transmitters Nor-adrenalin and Serotonin (5-HT), whose presence is presumably related to the non-olfactory innervations. Local stimulation effects [Figs. 5.2 and 5.5(a)] can alter the biogenic amine levels in the VNO of female mice, as a result of exposure to male conspecific urine, and consequent arousal of the suction-pump [c.f. Fig. 5.7(a)].

An additional and widespread neuroactive (transmitter-like) compound is nitric oxide (NO). This gaseous secretion is a product of the action of the enzyme NO-synthase on arginine. It is implicated in at least two roles: within the non-sensory tissues of the organ, and at particular synapses in the AOB. One nitric oxidergic effect is initiated by the nerve fibres supplying the smooth muscle component of the vasomotor tissues. The other effect is the expected action of NO on the output of the mucosal glands, in both the vomeronasal and posterior nasal sites (Kulkarni *et al.*, 1994). Centrally, NO is found in the mitral-granule cell layer, where it enhances the action of NorAd [Fig. 5.14(b)], as found in recently mated female mice (Okere *et al.*, 1996). The bulbar functions of NO in relation to reproductive events are somewhat paralleled by its action at higher centres concerned with LHRH production (Rettori *et al.*, 1993). Nitric oxidergic fibres assist with the production of prostaglandin E2, which in turn induces exocytosis of the releasing hormone from its bound state in storage granules. The free LHRH is then available for take-up by portal vessels to sustain basal pulsatile levels in male plasma. This terminal portion of the reproductive pathway is activated by AOS input, for instance in producing the androgen surges elicited by female chemosignals [Figs. 5.8(c) and 5.11].

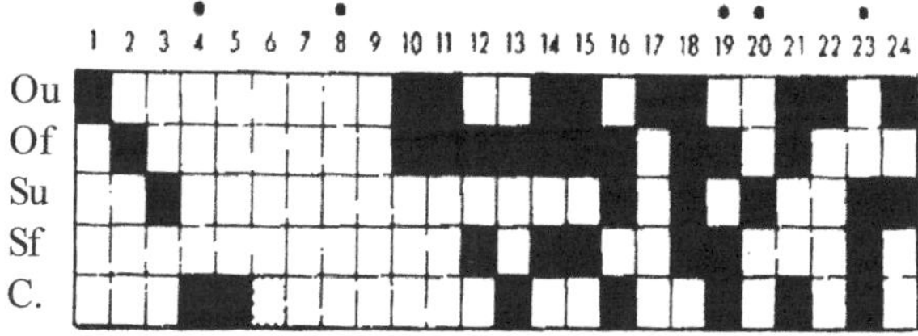

Fig. 5.5(b) Single unit responses in neocortex of dog: effect of N-P duct closure [= • columns] on response to conspecific odours. Ou = own urine, Of = own faeces, Su = strange urine, Sf = strange faeces, and C. = dry food for dog (from Onoda *et al.*, 1981).

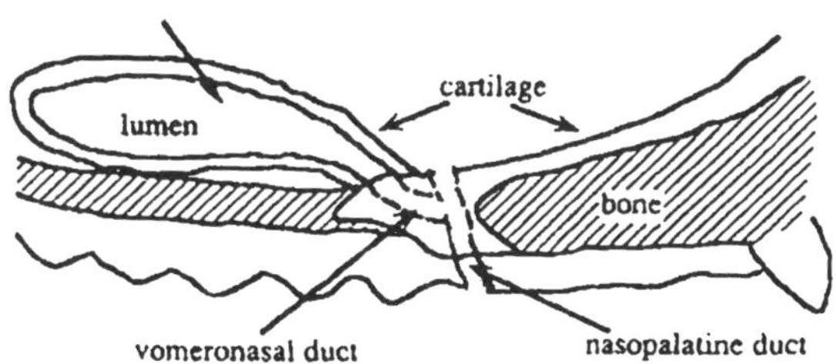

Fig. 5.6 "Deafferentation" of AOS by blockage of stimulus access. Prevention of VN output by injection of tissue cement seals lumen of (1) N-Pd (oral entry) and (2) VNd (nasal entry) (from Dorries *et al.*, 1997).

5.1.1 Stimulus Access

This is achieved by the operation of the organ's vasomotor mechanism via oral and possibly nasal routes. The nerve supply to the sinuses surrounding the organ derives from the nasopalatine nerve. Sectioning or stimulating these efferent fibres in hamsters abolishes or activates the lumenal flow into or out of the organ (Meredith and O'Connell, 1979; Meredith, 1982 and 1994). Similar results in the cat confirmed that sympathetic innervation of the VN complex regulates the action of the VN pump, even in species with the option of Flehmen-induced uptake (discussed in Chap. 7; and Eccles, 1982 and 1983). Female mouse urine elicits strong vasomotor contraction in males, consistent with lumenal retention of the stimulus (Fig. 5.7), since the suction effect through the N-P canal is sustained for up to 60 sec. (Hatanaka, 1992). Direct measurement of the intra-lumenal pressure changes shows that vasomotor activity is responsible for the observed alterations, i.e. sinus expansion precedes a pressure rise (Bland and Cottrell, 1989). Increase in blood volume produces compression on the organ's walls laterally, or circumferentially (Pl. 2D), in relation to the distribution of the sinusoidal network. Differences between the muscle layers of the

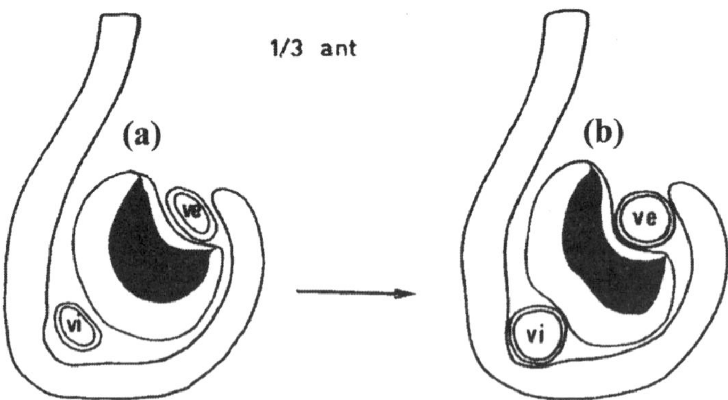

Fig. 5.7(a) Vomeronasal pump (vasomotor activation and control of VN fluid contents): reciprocal compression/relaxation of lumenal ■ and vascular volumes, vi/ve = internal/ external simus vessels. (a) Fluid intake — lumen expanded/vessels relaxed → pressure drops. (b) Fluid expulsion — lumen compressed/vessels expanded → pressure rise (from Schilling, 1970).

wall's vessels, medial and lateral to the organ, suggest that the operation of the vasculature is similar to that of other erectile tissues (Salazar *et al.*, 1997 and 1998). The vascular pressure is transferred inward towards the lumen, since the VN cartilage acts as a resistance container [Fig. 5.7(a)].

Tracer experiments have provided further evidence in support of the operation of a behavioural uptake mechanism and/or an autonomic pump in small- (mouse-lemur), medium- (guinea-pig), and large- (goat) bodied species (Ladewig *et al.*, 1980; Wysocki *et al.*, 1985; Schilling *et al.*, 1990). The behavioural effects (inhibition of mounting, mating) of interference with the pump were found to correspond to those of lesions to the organ's afferent output, since one prevents access to the receptor surface and the other to the brain (Murphy, 1980; Lehman and Winans, 1982; Meredith, 1982). The effector (NO) axons arise from cell bodies found outside the VN complex — in the sphenopalatine and trigeminal ganglia. The location of NO is supportive of its role in these regulatory mechanisms, since it is present in intraepithelial endings, but not within the receptors or supporting cells (Kishimoto *et al.*, 1993). NO fibres are concentrated in the lateral wall of the

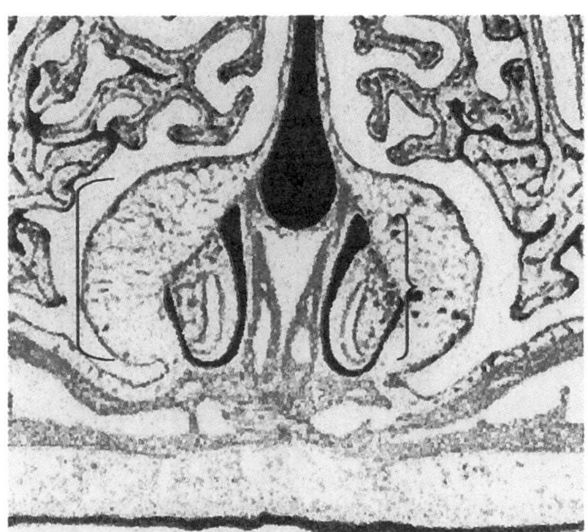

Fig. 5.7(b) Vascular Swell Bodies (SB): basal septal region in cat with bilateral cavernous tissue ([), enclosure of VN capsule (}) may allow alternation of pressure and modulation of VN lumen contents (see text) (after Negus, 1958).

main sinus; whether this localised distribution is correlated with that of the muscle fibres is conjectural at present (Soler and Suburo, 1998). Most of these motor neurones are also capable of producing the excitatory peptides VIP and Substance P (Matsuda *et al.*, 1996). An interaction between the peptides and nitric oxide in the control of tone in the cavernous vessels and in the secretion of mucous provides for a dual regulatory mechanism.

A previously unexplained observation of a link between nose and gonads — the so-called "naso-genital relationship" — can now be placed in context. Removal of the relay site for the pump-regulator (see above) by extirpation of the sphenopalatine ganglion was found to result in a state of pseudo-pregnancy (Rosen *et al.*, 1940). Interruption of stimulus access would clearly prevent most uptake through loss of control over the pump. A diestrous state would be induced (Sec. 5.3 below) attributable to the lack of male urinary semiochemicals.

Cavernous tissue sites outside the VN complex, but occasionally closely apposed to it, are the bilateral swell bodies. Their strategic location in the cat was illustrated by Negus [Fig. 5.7(b)]. Their placement on the base of the septum allows them to act as regulators of the left/right alternation in the respiratory airflow stream passing through the nose (Bojsen-Møller, 1971). As well as forcing inspired air into either one or other of the entrances to the maxillo-turbinal complex, the swell-bodies could provide a degree of compression on the VN complex (Eccles, pers. comm., 2001). The result of regulated bilateral switching of the input of volatiles was found to be a centrally-based improvement in the level of acuity achieved by the MOS (Sobel *et al.*, 1999a). By analogy, the alternation of pressure on the VN cartilage, its transfer to the organ's sinuses and then the lumen, might provide a similar enhancement of perception within the AOS. Experimental evidence for this hypothesis is lacking. The role of fluid ejection from the nose (Pl. 5.2) is temporally associated with the Flehmen cycle [Figs. 7.6(e) and (f)]. It is uncertain whether the expulsion by a "rinsing-out" movement of luminal fluid represents a preparation for sample intake, or a post-analysis removal of unwanted stimuli. The expelled liquid has not been analysed, or identified as definitively of VNO origin by tracer labelling. It is probably secretory, and is produced by VNCs with and

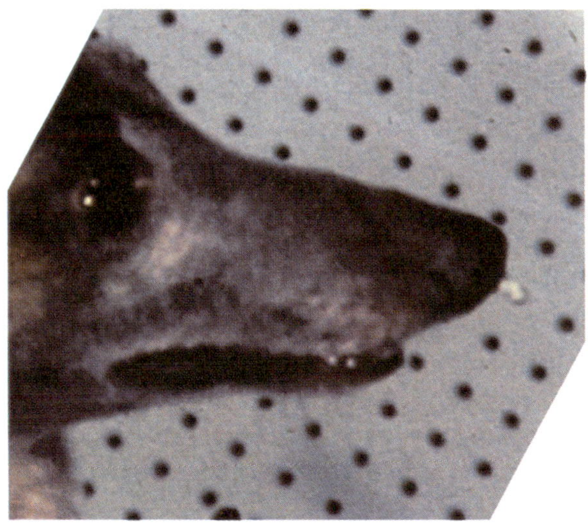

Pl. 5.2 Fluid expulsion: dog VN lumenal contents. Droplet forms at male nostril, during Flehmen response to estrous urine (courtesy, Fay Lindsay©).

without oral access (Jacobson, 1812; Lindsay and Burton, 1984; Lindsay pers. comm., 1985).

5.1.2 Comparative

5.1.2.1 *Fishes*

The EOG responses of bony fish, a shark and a cyclostome show marked sensitivities to several distinct water-soluble compounds: amino acids and bile acids, prostaglandins and steroids (Sorensen *et al.*, 1990). Goldfish (*Carassius auratus*) are differentially stimulated by l-amino acids and urinary steroids (Hanson, 1998). This could reflect the occurrence of "tuned" class I receptor sites whose genes show some resemblance to the VN2 genes (Freitag *et al.*, 1998; Speca *et al.*, 1999). The excreted steroids produce highly species-specific responses to a small range of pheromonal compounds (Sorensen, 1996). Bulbar recording in the goldfish demonstrated similarities in the distribution of excitatory, inhibitory or neutral responses to a range of signal compounds.

The anatomically distinct mitral cells and ruffed cells [Fig. 2.9(a)] show contrasting interactions. The resulting output induced a "drastic intensification of centrally transmitted information" (Zippel *et al.*, 2000). This type of central processing of reproductive stimuli allowed discrimination of preovulatory and ovulatory pheromones (c.f. Fig. 7.7). Individual compounds from some teleost urines activate discrete epithelial "hot-spots", or stimulus-specific neuronal clusters. Coalition of these into a single area may well be the precursor step in the segregation of receptor cells with either pheromonal or non-pheromonal responsiveness (Asano-Miyoshi *et al.*, 2000). At this level of tissue specialisation, there would then be at least the basic organisation for the emergence of the dual system of chemoreception. The high variance of form and function among fishes makes identification of the precursor steps in AOS differentiation somewhat problematic.

5.1.2.2 *Amphibia*

Common European frogs detect amino and carboxylic acids which induce AOB responses at 10^{-4} M, a sensitivity level comparable to fish (Kruzhalov, 1980). Simple proteins, such as the decapeptide sodefrin, stimulate the vomeronasal epithelium in sexually mature female newts (Kikuyama *et al.*, 1997). Its minimum effective concentration in water is 0.1–1.0 pmol/l, suggesting that high sensitivity is combined with the short-range transmission found in newt courtship [c.f. Fig. 7.2(a)]. The soliciting or female-attracting secretion from the abdominal gland of male Red-bellied newts evokes a marked EOG response in the epithelium of the lateral nasal sinus (LNS). The dose-dependent nature of the response from the accessory area supports the conclusion that it has achieved functional separation (Toyoda, 2000).

The lateral diverticulum cells in semi-terrestrial species such as toads can still detect a wide range of amino acids, comparable to the properties of fish neuroepithelium. Both water-soluble and volatile odourants are discriminated by the olfactory neurones of the Clawed toad (*Xenopus*) (Iida and Kashiwayanagi, 1999). When single olfactory neurones were tested with acidic, neutral and basic amino acids, over 50% of the receptors gave some excitatory response.

5.1.2.3 *Reptiles*

Both primitive and advanced forms have been examined for vomeronasal excitation. The turtle VN nerve was the subject of EOG recordings elicited by small organic molecules and by specific signal compounds (Tucker, 1963; Hatanaka, 1987). The simple arrangement of the chelonian accessory area (Fig. 2.8) allows air or liquid delivery; hence the preferred odourant vehicle varies with habitat across the aquatic or terrestrial turtles, and the land tortoises (Chap. 1; and Halpern, 1992).

The Squamates, as accessory specialists, acquire particulates by direct presentation to the oral entrances of their VN ducts (Chap. 7). A water-soluble 24 kDa glycoprotein is suggested as being the active component in prey (earthworm) washes; these elicit positive and negative single unit firing of AOB neurones (Meredith, 1978; Inouchi, 1989). Significantly, the organ was found to respond to liquid, but not to airborne delivery of "natural" stimuli; in contrast, no differential responses occurred to other standard odourants (Hatanaka, 1987). When partly anaesthetised, snakes did not show activity during tongue-flicks but only when prey extracts were presented to the VNO apertures; swabs pressed onto the upper oral surface elicited unit firing in the AOB (Meredith, 1978). When MOS or AOS input was removed by nerve section, tongue-flick (TF) rates and prey odour discrimination fell, or were abolished (Halpern *et al.*, 1997). Clearly both systems elicit stimulus uptake; airborne stimuli logically will alert the MOS first, and then activate AOS exploration in sequence (Graves, 1990). General arousal in novel environments is thought to stimulate non-specific chemosensory acquisition by TF, seemingly as part of low-level stress effects (Greenberg, 1993).

5.2 DEAFFERENTATION (MAMMALS)

The classical approach, advocated by the organ's discoverer (pp. 94), seeks to prevent any transmission along the afferent pathway, then to analyse the resulting deficits, if any. Adaptiveness in the AOS — such as its dependence on experiential variables — has also to be recognised and evaluated (Wysocki, 1986; Clancy *et al.*, 1988). The aim is to disentangle the various contributions of the AOS and MOS to the interpretation and perception of chemosensory information, plus selection of the appropriate response. Secondary anosmia by conventional surgery would ideally be replaced by non-invasive approaches which leave all components of the MOS and other nasal chemoreceptors intact. Interruptions of a specific sensory input should selectively remove responsiveness to a limited class of stimuli and not interfere with other pathways. This is rarely, if ever, achievable, especially in mammals with their multisensory capabilities and the parallel processes occurring within chemoreception.

The consequences of removal of the accessory bulb (AOB-x) should be scrutinised in relation to complete deafferentation (MOB-x), since if bulbectomy is the sole manipulation, little is provided of interpretative value (e.g. Baldwin, 1980; Kapusta, 1996). Partial intervention to remove the AOB is surgically fallible, whereas selective chemical destruction of the VNE and MOE are more feasible. Most investigations rely on verifiable interruptions along the pathway, from peripheral to terminal-central structures, post-operative evaluation being essential (Wysocki and Wysocki, 1995). Negative results (considered below) which seem to confound predictions of an AOS involvement are often as illuminating as those which support a functional association (Wysocki and Lepri, 1991). Stimulus uptake blockade by destruction of the VN-duct and/or the naso-palatine duct was one of the very first procedures applied (Mihalkovics, 1898). Cautery of the palatal ingress in cats did not disturb feeding — the then prevailing assumption being that the organ was a gustatory sense.

Prevention of access is the least intrusive method since it need not have any irreversible consequences for the afferent pathway. The common entrance to both olfactory systems in newts is easily closed-off by plugging the nostrils (Kikuyama *et al.*, 1997). A potentially reversible method threaded plugs into the NP canal of cats via the nasal cavity (Verberne, 1980). This procedure produced a slight effect on male chemoinvestigation of urine and or scent marks. The advantages of avoiding tissue disturbance then, have to be offset by the lack of any estimate of the effectiveness of the blockade, especially if reversible. Tissue cement injections into the N-Pd can be applied to the larger

(>5 kg) mammals, and are effective in preventing uptake of the male pig chemosignal androstenone (c.f. Fig. 5.6; and Dorries *et al.*, 1997).

An alternative approach is to examine the onset of patency of the duct lumens from mouth to nose. Developmental changes in functional onset can thus be related to the establishment of a through route for fluids [Chap. 4 1(c)]. Mechanical (microsphere) tracers have been employed with success to fix the timing of access to the VN lumen in relation to birth, and to indicate whether and when amniotic fluid contents can reach the VN lumen (Coppola, 1993; Coppola and Millar, 1994). Similarly, to establish whether licking or rhinarial contact does result in fluid uptake (Chap. 2.1), chemical tracers (fluorescent, radio-opaque, etc.) have been added to urine and successfully detected in the lumen after chemoinvestigation (Hart, 1983; Wysocki, 1985; Schilling, 1990). Complete and selective removal of the oral intake route in rodents demonstrates that nasal access to the organ is unimpaired. Destruction of the male hamster N-P duct does not influence sexual responses (Meredith, 1991b). Possibly some nasal entry, channelled by the sub-septal groove [Fig. 2.4(c)], is sufficient to reach the separate VN duct.

Amongst reptiles there are multiple chemoreceptive usages for the tongue; the most important being stimulus-sampling and uptake (Graves, 1990). Flicking of the forked tongue by lizards and snakes was long suspected of involvement in odour molecule transfer [Fig. 7.3(a); and Burghardt; 1979]. VN uptake of a labelled amino acid (3^{Θ}Proline) by a lizard was (predictably) prevented by tongue removal. However, nose contact alone did allow stimulus access, provided the lizard was allowed to touch a swab soaked in prey-extract, as in palatal uptake by snakes (Meredith and Burghart, 1978). Closure of the VN entrance by tissue adhesive prevented uptake of label under both conditions (Graves and Halpern, 1989). The effectiveness of snout contact alone is unexpected in this group. The presence of a "lingual notch" in some snakes suggests that a degree of oral uptake might occur without tongue extrusion through the V-gap formed by the closed jaws.

Extirpation of the entire VN complex (VN-x) is an effective and informative approach applied largely to mammals. A palatal incision, followed by cautery or dissection of the capsule, usually removes all sensory elements and should produce about 95% degeneration of the glomerular layer in the AOB (Wysocki and Wysocki, 1995). Partial ablation will often confound functional tests, and in rodents and lagomorphs the separation of the two ducts [Figs. 2.4(a) and (b)] implies that removal of the N-P duct alone is insufficient to prevent stimulus

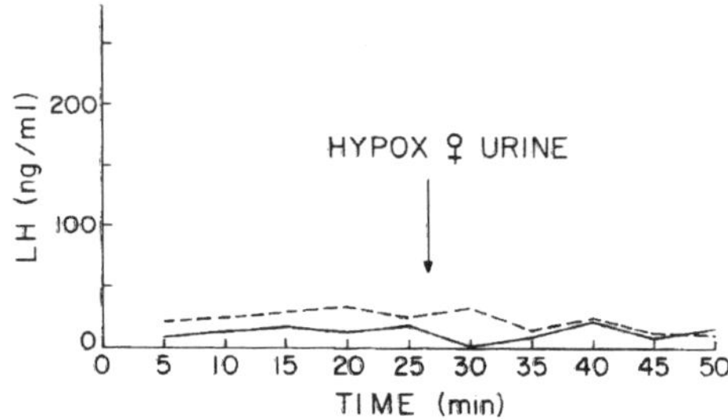

Fig. 5.8(a) Hypophysectomy abolishes urinary stimulation: ablation of anterior pituitary removes neurocrine linkage via AOS to testes; --- = before; — = after operation (from Johnston, 1985).

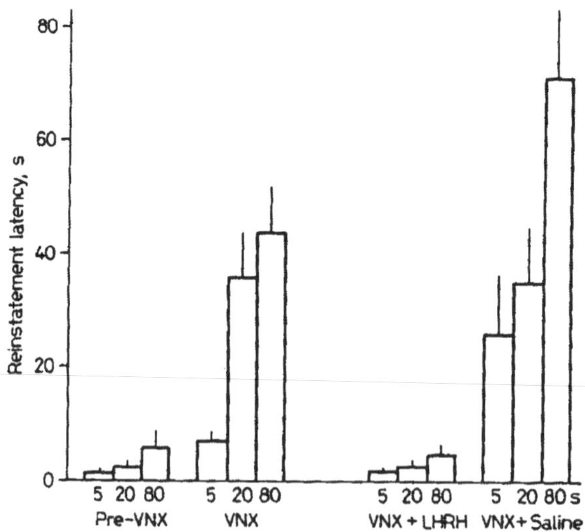

Fig. 5.8(b) VN-x and releasing hormone effects on female receptive behaviour: facilitation by LHRH and latency to tactile induction of lordosis in hamster (latency duration, sec.). LHRH restores responsiveness over saline control (from Mackay-Sim and Rose, 1986).

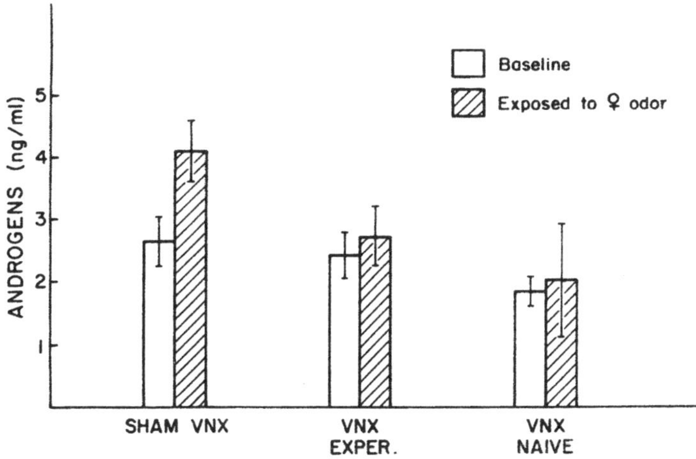

Fig. 5.8(c) VN-x male hamsters and inhibition of T.-response; induced by vaginal secretions, experienced and inexperienced Ss. (from Johnston, 1992).

access (Meredith, 1991b). Recovery time is a further confound, since if prolonged, it may allow regeneration and thus partial recovery of function may be present (c.f. Chap. 4). Transection of the VN nerve fibres as they cross the septum, but before they pass through the cribriform plate, is complicated by the dispersed nature of the axonal fan crossing the septal mucosa (Figs. 2.9 and 2.10).

An early effort using this approach in Guinea-pigs reported variable results — attributable to incomplete nerve sectioning (Planel, 1953). Sectioning procedures may also produce some unwanted effects as intracranial nerve section will remove part of the animal's *N. terminalis* sensory capability (Devitsina and Cherova, 1992).

The equivalent procedure for the main olfactory epithelial sheet (MOEx) is chemical ablation by treatment with (5%) zinc sulphate solution. The adoption of this approach was determined by practical anatomy, given the inevitably partial, let alone traumatic, results of cutting/scraping epithelia from the intricately folded sensory surfaces (Negus, 1958).

Removal of the AOB is handicapped by its inaccessibility and lack of external demarcation, making it a challenging target for complete removal (Cooper, 1974; Kelche and Aron, 1984; Dudley and Moss,

1994). When combined with VN-x it provides a useful parallel approach, as AOB-x leaves other nasal afferents untouched. However, in some species it may also damage MOB efferents (Beltramino and Taleisnik, 1983; Meredith, 1988). Systematic and sequential deafferentation by VN nerve section can yield additional information by examination of the consequences for local degeneration in the AOB (Roland *et al.*, 1995).

Central transection of the output LOT fibres from AOB to the higher brain centres, and destruction of functional projection sites within the cortex have been increasingly attempted (Raisman, 1972; Rajendren and Dominic, 1986; Demas *et al.*, 1997). These rely on accurate stereoscopy of the brain, hence this approach requires an expansion of the range of detailed 3D atlases available for domestic and non-domestic species (e.g. Bons *et al.*, 1998; Felix *et al.*, 1999).

At the sub-cellular level, genome manipulations include deletion of sequences from embryonic stem cells, and the construction of knock-out (KO) mutants. OMP-deficient mice (OMP-null mutants) show up to 40% reduction in EOG responses (Buiakova *et al.*, 1996). The suckling of mouse pups with a Golf deletion fails in 75% of mutants (Belluscio *et al.*, 1998). The anosmia which rendered the MOE ineffective may have extended to the comparable VN alleles, since pups are strongly oriented by a VN signal (Chap. 4). Expression of the dual OR genome can now be dissected by removing unique AOS and MOS components, as in comparison of anosmias due to adenylyl cyclase (ACIII.) deletions with those of IP3 deletions. The degree of divergence within chemoreception will remain incompletely known until such analyses are applied to all functional steps (Wong *et al.*, 2000).

5.3 NEUROENDOCRINOLOGY

5.3.1 Chemoinvestigation

Inexperienced males deprived of the VNO prior to mating do not respond on initial exposure to females — they do not vocalise or show T. surges (Wysocki, 1983). Androgen output in these conditions is not affected by the MOS or by visual cues, suggesting that in naïve mice the AOS-mediated chemosignals are processed unmodified, and so reach the anterior pituitary and induce LH secretion. The effects of VNO loss for males before any post-weaning experience with females are thus more drastic than loss following heterosexual encounters. Provided males are already fully sexually competent, then they can retain mating protocols and compensate for the absence of VNO input (Meredith, 1986). Hormonal responsiveness is thus maintained in VN-x and experienced, animals. Removal of the male VNO prior to any

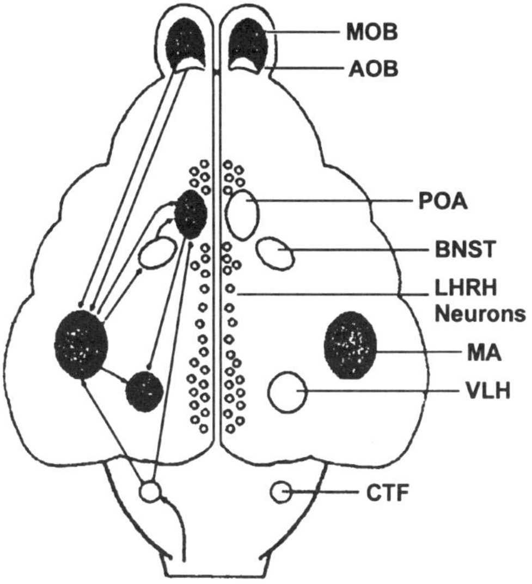

Fig. 5.9(a) Sexually dimorphic processing in Ferret: differential activation of central nuclei by heterosexual cues. Fos-induction levels: • moderate/high, L — female, and R — male, brains (from Wesinger and Baum, 1997).

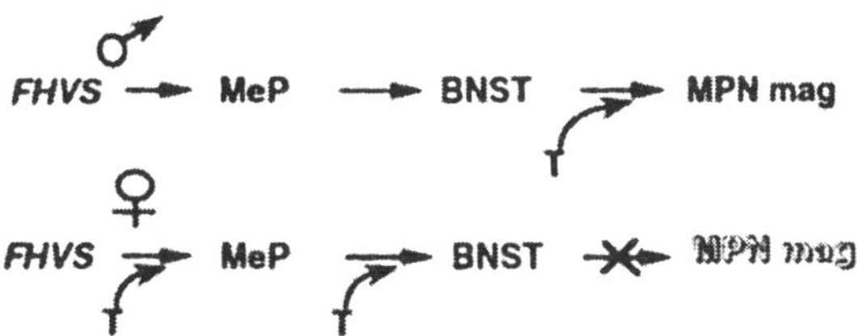

Fig. 5.9(b) Sex differences in responsiveness to female hamster vaginal fluid (FHVS): androgen (T) effects on central transmission pathways (from Swann and Fiber, 1997).

interaction with the female, removes mating patterns in several other rodent species.

Adult male Prairie voles performed with reduced levels of chemo-investigation to females after VN-x and did not complete successful insemination over an eight-week period (Wekesa and Lepri, 1994). The disruption of mating behaviour in mice is expressed through fewer mounting attempts; during these the frequency of intromissions and ejaculation was significantly lowered in lesioned males (Clancy *et al.*, 1984). The hamster falls into an intermediate category regarding mating. VN nerve cuts abolished sexual behaviour in 44% of males; the remaining hamsters, possibly with lower thresholds for female stimuli, showed mating persistence until the MOE was ablated (Winans and Powers, 1977).

Hormonal influences may be limited to sexually relevant cues, since not all scent marks are socially relevant in all situations (Petrulis *et al.*, 1995 and 1997). Chemoinvestigation by male hamsters of the female-indicator compound DMDS was independent of T. and did not differ from that of females. In contrast, the frequency of chemoinvestigation by castrates to vaginal secretion (containing DMDS) was enhanced by T.; intact males investigated FHVS five times more than they do the females.

Loss of proceptive behaviour also follows VN-x in female rats, in which the receptive posture (lordosis) is almost abolished (Saito and Moltz, 1986). Estrous cycles with receptivity were still maintained, since males given extended access to VN-x females inseminate them successfully.

That the mechanisms of intersexual recognition in mammals could be mediated wholly or partly by scent, occasioned some early speculation (Steinach, 1894). The most likely communication vehicle for females to indicate their estrous state was identified as urine (Beach and Gilmore, 1949). Typically, the findings from many mammals show that cyclic changes in male olfactory responsiveness to urine occur in response to variation in signal content, in turn reflecting the phases of ovarian activity (Blissitt *et al.*, 1990). The involvement of the AOS as the primary sensor was suspected from the association of urine-tasting with pre-mating responses (Estes, 1972). Vomeronasal uptake was established through the responsiveness of the AOS to reproductively important chemosignal sources (Wysocki *et al.*, 1983). Chemoinvestigation by male VN-x guinea pigs was not maintained; female-oriented responses being suppressed for up to four months post-operatively. The decline and loss of responsiveness was attributed to the inability of the MOS alone to maintain arousal. The extinction of the response may occur due to the loss of accessory reinforcement of the central mechanism supporting the behaviour (Beauchamp *et al.*, 1982 and 1985; and Chap. 7). Proceptive (courtship) responses by male mice to females or their urine include high frequency (70 KHz) calls (Dizinno *et al.*, 1978; Nyby *et al.*, 1983). This chemically induced aural output is preferentially mediated by the VNO in mice and rats (Chap. 7; and Bean, 1982; White *et al.*, 1991). The utility of combined chemo-auditory cues is explicable by the males' responses to chemosignals from novel females. These potentiate sequential mating (Coolidge effect), since testosterone levels in plasma rise sharply in response to female chemosignals, provided the VNO is intact (Macrides *et al.*, 1974; Purvis and Haynes, 1978; Wysocki *et al.*, 1983).

5.3.2 Photoperiod Effects

Prairie voles a monogamous, but group-dwelling species with spontaneous estrous, contrast with the polygynous but solitary meadow vole (*M. pennsylvaticus*). In the meadow vole, there is little effect on mating when females are kept under winter conditions of short photoperiods (10L:14D; Meek *et al.*, 1994). Under summer conditions (14L:10D), the overall percentage of females mating and with short latencies is unaffected by VN-x. The dominance of photoperiods over chemostimulation is clearly advantageous where the social system is adapted to multiple mating. Pair-bonded species such as the Prairie vole, perhaps require inter-partner stimulation to coordinate reproduction. Species differences in photoperiod effects are more likely than not, since hamsters, whether under short-day (SD), or long-day exposure, were unaffected by VNO loss (Pieper *et al.*, 1989). Central, not peripheral, deafferentation is more effective on the light response. In the male, sensitivity to SD-photoperiods shows as testicular regression; disinhibition of this response occurs after post-bulbar section of the lateral (LOT) tracts.

The AOS contributes elements to reproductive systems later in the life cycle, but these are dependent on species requirements in the extent of their influence. A model of socioecological interactions is further discussed in Chap. 7.3.

Chemoinvestigation patterns in female hamsters are maintained even though part of their amygdala — the medial zone — was removed (Petrulis and Johnston, 1999). They were still able to discriminate between scents of individual males, suggesting prior AOB processing. When given a Y-maze (two-choice) test for discrimination, no preference for male versus female odours was displayed. Conspecific odours did not elicit scent deposition, since over-marking with vaginal or flank gland secretions was reduced or eliminated. It is tempting to interpret such results with reference to the functions of the medial amygdala as a cross-over or exchange site for chemosensory information. Since input via both the MOS and the AOS pathways is coincident in such a restricted neural site [Fig. 5.15(b)], it has been proposed as a convergence point or "node" for chemosensory information (Licht and Meredith, 1987; Meredith, 1998). Direct effects on sexual behaviour mediated by the medial amygdala (mAMYG) are evident in the receptive posture of female rats. In ovariectomised rats, effects on lordosis were evident ten days post-operation and lasted from up to 50 days (Dudley and Moss, 1994). The elicitation of lordosis in hamsters is also affected by loss of the VNO [Fig. 5.8(b)]. The inhibitory effect is on latency to expression, and is lifted by LHRH (Mackay-Sim and Rose, 1986).

The importance of the relay functions of the pathway is supported by lesions which remove key nuclei that transmit sex behaviour input processed by the AOB (Lehman and Winans, 1982). The rostral part of the cortico-medial amygdala has more effect on male hamster mating than VN-x. The integration of volatile or non-volatile inputs is a key variable for some of the social discriminations processed at this level. Chemical (NMA) lesioning of amygdala to selectively damage the cortico-medial nucleus results in decreased paternal behaviour in Prairie voles. These neurones appear to be essential for maintaining "partner" female contact, as well as responses to offspring. Selective processing

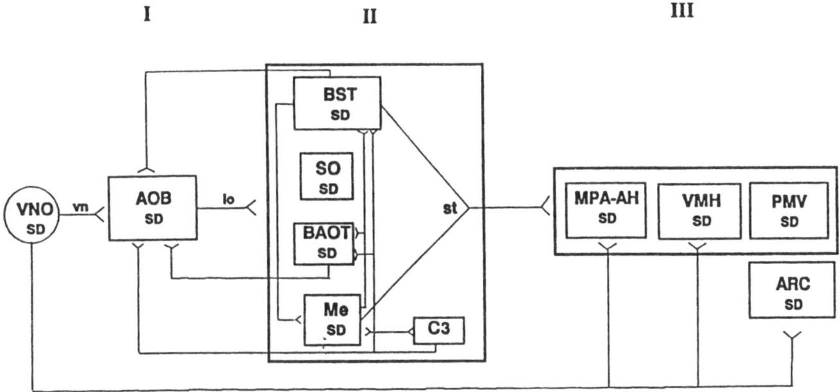

Fig. 5.10 Sexually Dimorphic Network: Structural and/or functional distinctions at peripheral to central levels (I to III) in the rat AOS (from Segovia and Guillamon, 1993).

of sex-related stimuli occurs in the premamillary nucleus (PMv) of male mice. Exposure to female-soiled bedding induced Fos-ir expression only in this nucleus, not in other chemosensory nuclei of the amygdala or hypothalamus (Yokosuka *et al.*, 1999).

Brennan (1995) found that female mitral cells are capable of distinguishing the input arising from male signals by a sustained dis-inhibition; re-exposure to stimuli from the same male is potentiated over a two-day period. Whereas, following mating the usual responsiveness noted in the Bruce effect (p. 123) is the enhanced inhibition mediated by GABA. The chemosignal effect of male urine involves a rise in the excitatory transmitters Glu and Asp at the reciprocal synapses between mitral/tufted cells [Figs. 5.12(a) and (b)]. Male hamsters exposed to urine from female conspecifics, and from rat females, showed alterations in synaptic morphology which could provide the basis for the selective enhancement of sex signal processing (Matsuoka *et al.*, 1998). As seen in female mice, the initial excitatory synapses at mitral/tufted cells were potentiated by increases in size. Observable structural expansion is presumably directly associated with the enhancement of transmitter production. The inhibitory symmetrical synapses did not show any such change following exposure to heterospecific urine. Confirmation of functional linkage between chemosignal processing and adaptive synaptic alterations is desirable for other semiochemical types.

5.3.3 Sexual Dimorphisms

Several lines of evidence obtained from the AOS of mice and rats support the contention that the AOS displays a sexually distinct morphology at cellular and sub-cellular levels, and in associated physiological responses (Simerly, 1990).

In rats, dimorphisms are reported in cell size and density in the presence of steroid receptors (androgens and estrogens) and the related differential sensitivities to gonadal hormones on central neurones (Segovia, 1999). The proposed mechanism invokes the immediate post-natal (in rodents) differentiation of the brain which determines sex-oriented structures and later responses to sex-typical hormones (Chap. 4; and Cooke *et al.*, 1998). The VNO itself has not been systematically examined for all relative male/female distinctions. The volume of the organ in each sex relates to body size. Hence, without reference to overall growth rates, sex differences in tissues are not entirely reliable biological indicators (Weiler, 1999). However, most structures that receive vomeronasal input do show some or all of the dimorphisms so far described (Segovia, 1984). Female AOS neurones are present in lower numbers and have a smaller cell volume than those of male rats. Males have larger post-bulbar units in the medial preoptic area, and several nuclei in the AOS pathway: the bed nucleus of the accessory olfactory tract (BAOT), and of the stria terminalis (BST), the medial amygdaloid, ventromedial hypothalamic, and the ventral region of the premamillary nucleus (Segovia, 1993). Although these nuclei appear mostly excitatory during transmission along the AOS pathway, opposite effects occur. From the bed nucleus of the accessory olfactory tract (BAOT), an output exerts a persistent (tonic) inhibitory effect on parental behaviour in all males and in virgin female rats. Removal of this influence allows lesioned males to express maternal responses to pups (Izquierdo *et al.*, 1992).

It should be noted that the central nuclei of the MOS pathway are also differentially sensitive to steroids, and e.g. selectively process stimuli arising from female volatiles in the hamster and ferret (Swann, 1997; Wesinger, 1997; Kelliher, 1998).

Findings on functional sexual distinctions in the AOS of rodents support the structural evidence. The BNST and MPOA of female mice are differentially activated by male soiled bedding stimuli, whereas males themselves did not respond in this manner (Halem, 1999). In the same study, sexually naïve ovariectomised (Ov-x) mice (Balb/c strain) were treated with estradiol and exposed to male stimuli. Their VNO contained significantly more active neurones than controls of each sex. A design using direct recording from the male neuroepithelium extended an earlier finding that particular types of VN neurones were "sex-tuned" to urinary stimuli, as seen [Fig. 5.3(b)] in the burst of urine-induced firing (Hatanaka *et al.*, 1992; Trotier *et al.*, 1998). The VN epithelium (Fig. 5.12) contained a mix of uniquely male-responsive neurones co-existing with the alternate "female" receptor cells (Holy *et al.*, 2000). Their occurrence was approximately equally balanced; ~50% of VNORs can be activated (Fig. 5.11) by either male or female urine (at 300 × dilution). The threshold sensitivities of the unisexual responses in male/female cells differed by up to 10^3.

The accessory system in male and female rodents then, is susceptible to a variety of modulatory influences, which in turn affects the functions of the reproductive system. The interactions of hormones with transmitters at each stage along the AOS pathway (Fig. 5.9) form the

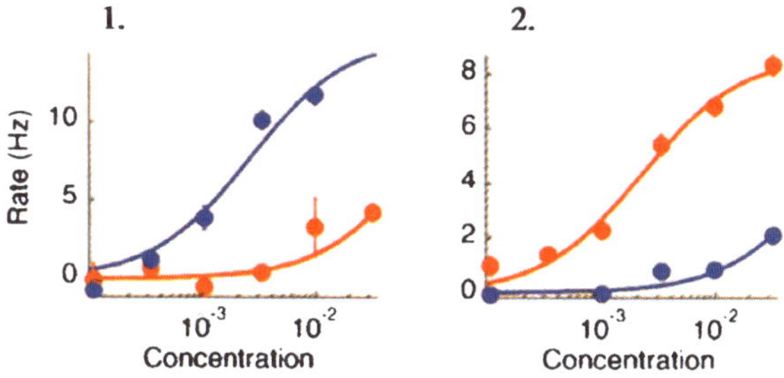

Fig. 5.11 Output rate from VN epithelia in male mice: sexually differentiated responses to male **(1.)** and to female **(2.)** urine, [time = sec]; ~50% neurones preferentially respond to like- or to opposite-sex urine (@300 × dilution) (from Holy *et al.*, 2000).

basis for attempts to explain the process of integration of internal with external signalling.

Sexually differentiated responses to female-soiled bedding occur within the regions of the AOB; only the rostral zone was *fos*-activated in males as opposed to that of females; the caudal zone cells had no such differential activation; VN-x removed the response (Dudley and Moss, 1999; Matsuoka *et al.*, 1999). The number of *Fos*-ir cells was larger after exposure to females of ICR stain, than to BALB females. Male strain-differences gave equivalent amounts of rostral/caudal activity in females. It is probable that discrimination by females of between-strain chemosignals is related to the fine distinctions possible through the operation of cues related to influence of products from the MHC loci.

Regional AOB effects are found in female rats when they are exposed to intact urine from males; the rostral *fos* levels being twice as high as caudal ones. Female urine gave similar but less marked responsiveness in the main bulbar zones but discriminatory (caudal > rostral) responses were located in lateral regions (Inamura *et al.*, 1999). The VN input shows some patterning of the initial bulb responses, indicating that some location-specific processing may aid further discrimination of urinary stimuli. Activity of more central regions of the AOS pathway reveals that some of the anatomical and histochemical sexual dimorphisms are accompanied by functional distinctions. Stimuli from males, and from appropriate gonadal steroids, probably combine to alter the properties of the VN, bulbar and cortical neurones (Wood, 1995).

The properties of the central regions are such that they can still respond to endogenous hormone(s) in the absence of VN input. The amygdala or other central regions are so organised that mating responses can be elicited by appropriate combinations of stimuli. Most, if not all, of these effects may arise from the influence of the local action of steroid-sensitive neurones on other sensory pathway cells [Figs. 5.9(b) and 5.10]. The facilitation by and effectiveness of sex steroids arises from their ability to increase the overall efficiency of transmission; steroidal effects being mediated by increases in neurogenesis, dendritic branching and synapse density, plus upregulation (activation) of transmitter synthesis. Multiple facilitation of this kind enables the central parts of the AOS to respond more rapidly to already filtered stimuli which arrive from the accessory bulb. The central areas are thought to compensate for loss of input by raising the activity of hypothalamic cells. These in turn respond as non-limiting or passive elements; local injection of transmitters produced advanced puberty in macaques (Plant, 1989).

Hence, it appears that central actions of the releasing hormone itself, not via pituitary hormone effects, are responsible for the ability of the system to maintain mating behaviours in the absence of the normal sensory cues. Increased LHRH capacity occurred after bulbectomy (Ichikawa and Oka, 1988). Intra-cerebral LHRH injected into the ventricles of VN-x, and inexperienced, males relieved their mating behaviour deficits, as did tests with a similar but non-LH releasing analogue (Fernandez-Fewell, 1995). This finding suggests that the LHRH peptide itself may facilitate male mating behaviour through an extra-pituitary route. It further implies that the neural circuits mediating mating might become established by early releasing-hormone delivery to the developing brain. The lack of AOS input hinders subsequent release in

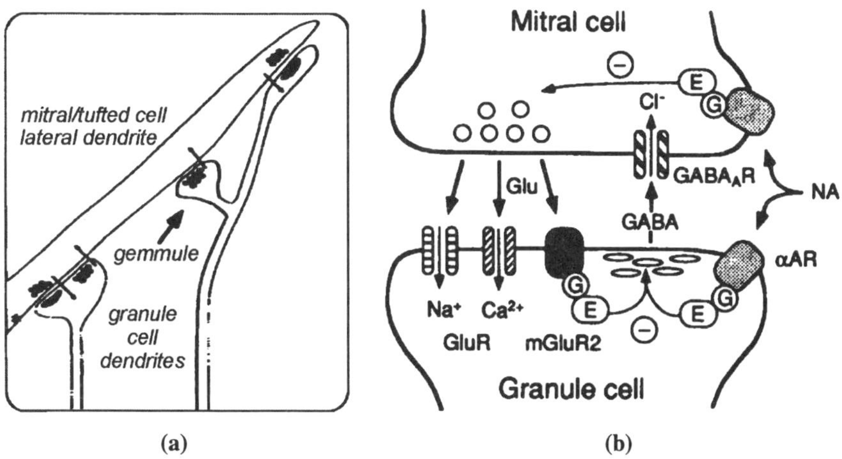

Fig. 5.12 (**a**) Synaptic types along dendritic spines of M/T and GC units; uni-, and bi-directional junctions. (**b**) Transmitter systems at a reciprocal synapse, Mitral–Granule cell junction. [Glu, glutamate (R, receptor); GABA, γ-aminobutyric acid (R, receptor); E, intracellular effector; and αAR, alpha-adrenergic receptor.]. (From Hayashi *et al.*, 1993.)

the absence of any compensatory learnt responses associated with other stimuli. Neuropeptide injection short-circuits this deficit in adults to act upon the already primed central neurones. The fertility of both sexes is substantially regulated through the direct access of their AOS to the neuronal and neuroendocrine cells which regulate gonadal functions.

5.3.4 Sociosexual Effects

One model for neurocrine control over cyclicity in rodents proposes a coupling mechanism linking semiochemical input with cycle phase (McClintock, 1983). Signals of male or female origin initiate a set of hormonal changes responsible for the initiation, increase or decrease, suspension and synchrony of gonadal function(s). Several "named" effects, discovered and analysed with laboratory strains, are particularly relevant to the AOS. Whether some or all of these are genuinely adaptive for members of wild populations (c.f. Chap. 7.3) is not fully established (Chipman and Fox, 1966; Bronson, 1989; Heske and Nelson, 1989). However, feral Deer mice (*Peromyscus* spp.) display (in captivity) several intriguing reproductive responses of this type, such as cycle-synchrony and implantation failure (Bronson, 1964; Haigh *et al.*, 1988). Nevertheless, researches on inbred rodents established the primary physiological role of accessory olfaction, and contributed substantial insights on neurocrine adjustments to social conditions.

The absence of any exposure to male odour is alone responsible for vomeronasal-mediated effects on grouped females. Inter-female sociostressful conditions (the Lee-Boot Effect) suppress estrus, whilst the induction of cyclicity in such anestrous grouped females (Whitten Effect) is attributed to exposure to male urine (von der Lee, 1955; Whitten, 1956). The acceleration of puberty by advancement of the start of cyclicity, further extended the range of male-associated effects (Vandenbergh, 1967; Terman, 1984). Similar effects including pubertal inhibition, and cycle synchrony are also attributable to male (or female) chemosignalling (McClintock, 1983). Their existence in humans while intriguing, has necessitated an extensive and critical analysis of the problematic causal influences (Preti and Wysocki, 1999). Despite difficulties inherent in human studies, limited support for the

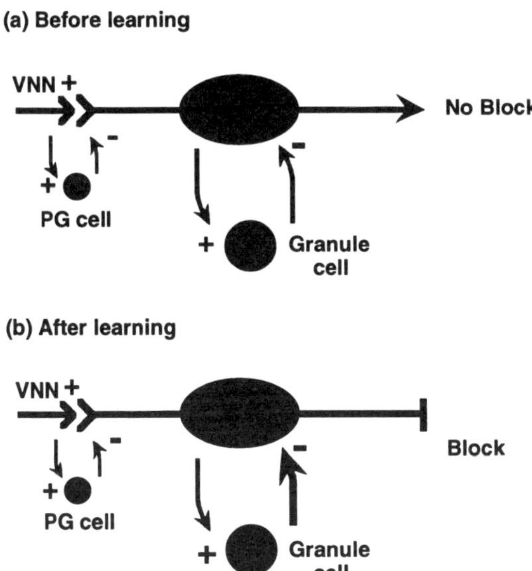

Fig. 5.13 Pregnancy Blockade, synaptic changes underlying processing of male chemosignals: female mouse AOB local circuits before (**a**), and after (**b**) mating. Transmission (gating) follows exposure to strange male odours (from Brennan and Keverne, in press).

existence of cycle-synchrony has emerged (Welder, 1993). Amongst the possible causal factors of variability in man is the occurrence of anosmias, such as those affecting axillary signals (Preti *et al.*, 1995; Wysocki *et al.*, 1999).

Another related, apparently dysgenic effect, has received sustained analysis as a model of localised memory induction. The "strange-male" or pregnancy-block (PB) effect, is named for its discoverer Hilda Bruce (Bruce, 1959). It is the premature curtailment of a conceptus by exposure to urinary chemosignals from a non-paternal male (Hoppe *et al.*, 1975). Interactions during mating allow AOS uptake of urinary compounds which will later serve to identify the presence of the stud male. The protein constituents of urine (MUP, Chap. 3) are implicated as part of the "male-signal" complex in PB (Marchlewska-Koj, 1981). The correct identification of the known versus unknown male occurs provided the female AOS is given a four-hour period in which to register an olfactory profile in bulbar circuits. This individual signature can be recalled as a memory trace to "match" the initial mating chemosignal profile against that of others. Interruption of this process immediately after mating can prevent formation of an effective working memory of the stud male. If the presence of the stud male is detected, or that of other females, then the effect is abolished.

Male signals are perceived in the female's AOB; if no recognition of the stud male's chemosignal profile can be made, then transmission block to the hypothalamus is lifted (disinhibition). The consequences for the female are the interruption of her pre-implantation phase, since the relaying of messages from the bulb sets off an irreversible train of neural/neurocrine events. The effector route [Figs. 4.4(b) and 5.8] via the anterior pituitary cumulates in the loss of the ovarian hormones required to maintain the pregnancy. The end point is the failure of the corpus luteum and of its secretion; the luteal steroid progesterone being needed immediately post-coitum. A strange-male exposed female given progesterone is thus protected from failure, and is able to continue with her pregnancy, but only if she receives it during the first two days following mating (Vandenbergh, 1975; Marchlewska-Koj, 1981; Rajendren, 1993).

Analysis of the neural chain of PB events by Keverne and colleagues has disentangled an event-specific "local memory" circuit which allows the female to abort her pregnancy (Keverne, 1990 and 1996). This ability relies on the filtering capacity at the level of the AOB alone (Kaba *et al.*, 1992). The input from the organ arrives at the 1st (glomerular) synapse and activates the main dendrites of one type of mitral/tufted cells (M/TC). The AOB glomeruli, like those of the MOB (Chap. 2) are functional modules, but incorporating clusters of fibres from two or more cell types (Figs. 5.14 and 5.15). At the initial, uni-directional synapse, stimuli arrive from the axons of only one type of OR or VNOR, in any one glomerulus. The distinction between the systems lies with the pattern of interconnections which represent the first processing step for incoming sensation. Models representing coding patterns for both systems are set out in Figs. 5.14(a) and (b). The synapse on the M/TC dendrite which may or may not allow transmission out of the cell is a reciprocal or two-way junction [Fig. 5.12(b)], with a granule cell

(GC) axon (Jia *et al.*, 1999). The GC are the main inhibitory interneurones, while the peri-glomerular cells can alter the probability of transmission at the first synapse. The olfactory inputs to the M/TCs have two ways in which they may be connected, via their primary dendrites, to a particular glomerulus. First, they may supply only one functional type as in MOE input [Fig. 5.14(a)]. Second, they may supply two or more functional types [Fig. 5.14(b)]. The single connectivity type is found in the MOB, as the primary OR–M/TC

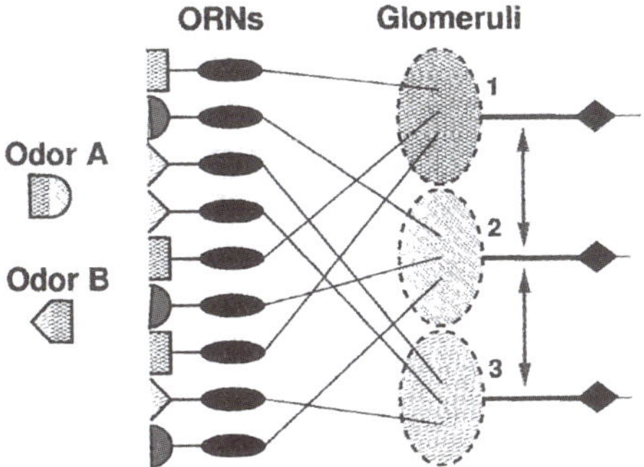

Fig. 5.14(a) Coding of MOS information within main bulb: OR neurones of same receptor type e.g. ▭ converge on a single MOB glomerulus; pattern-generation by integration of input via two (or more) receptor cell types, e.g. odourant-binding to ORs A and B = activation of glomeruli 1 and 2 (from Christensen and White, 2000).

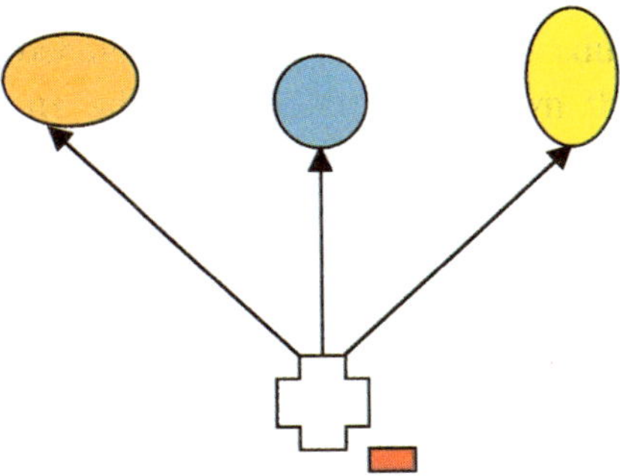

Fig. 5.14(b) Coding of VNOS information: VNOR neurones of same type diverge to several distinct Glomeruli. Input from a Vomodour ▬ reaches the AOB zone(s) in multiple projection(s).

relationship; the second, or multiple pattern, allows a M/TC to access input from several VNORs. This multiple type is specific to the either zone of the AOB, and is thought to provide flexibility of response and ability to diversify the way the AOS deals with the processing of

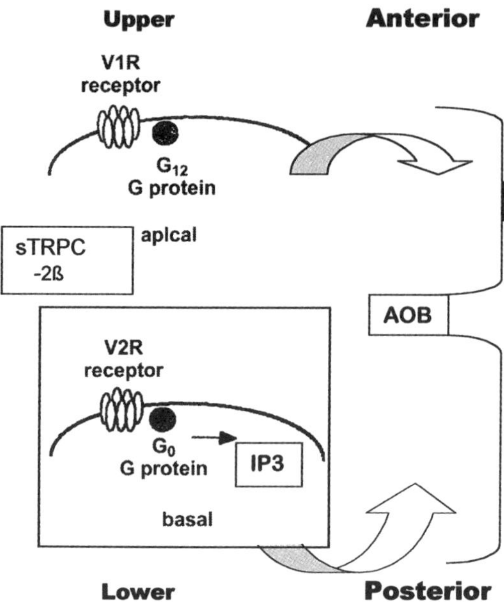

Fig. 5.15(a) Heterogeneity in rodent AOS: differential output from the regions of the VNE (upper/lower neurones, L.) to distinct regions of AOB, R.; apical ⇒ anterior; basal ⇒ posterior. Putative ion-channel (sTRP2) and 2nd messenger (IP-3) mechanisms.

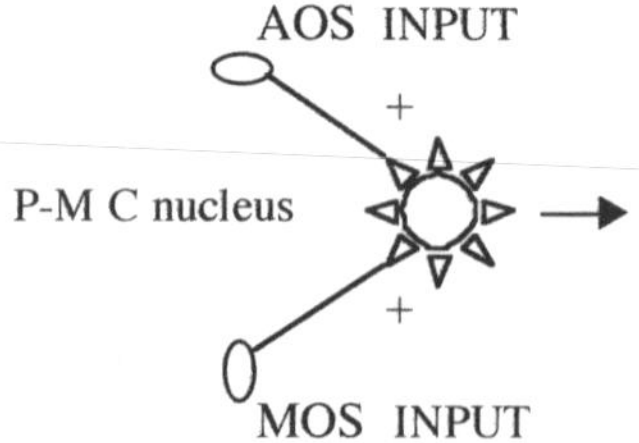

Fig. 5.15(b) Central integration of chemosignals: convergence of excitatory (+) stimuli on a single unit in the postero-medial nucleus of Amygdala; either input facilitates output → (Licht and Meredith, 1987).

vomodours (Brennan *et al.*, 1995; Brennan and Keverne, 2000 and in press). The arrangement may well reflect the differential connectivity of the VNOR regions to distinct bulbar zones [Fig. 5.15(a)]; complex signals being of necessity processed by relatively small numbers of AOS pathway neurones.

The mode of operation of the key M/TC–GC synapse which utilises Glu, GABA and NorAd to regulate output transmission has been finely analysed as the basis for the AOB's local memory. The process is summarised in Fig. 5.13; it involves on/off switching of an excitatory pathway for male urinary stimuli. Arousal via intromittent or other stimuli at or before mating increases NorAd levels at the reciprocal synapse, due to activation by the centrifugal axon terminals [Figs. 5.4, 5.12(a) and (b)]. The M/TC which respond to the first (stud) male stimuli are potentiated by this adrenergic effect. At the same time, they also start up the reverse inhibitory action of the synapse. This occurs in mated, but not in non-mated females, probably due to a Glu-synaptic enhancement (Brennan, 1995; Matsuoka, 1998). As noted above, adrenergic actions are NO mediated, since, if prevented from acting by inhibitors, the local memory representing the stud male fails to be established (Okere, 1996). The increased inhibitory GC action prevents transmission, but only of recognised stimuli, hence continuance of the pregnancy initiated by that male. A second male provides a distinct input pattern reaching a separate set of non-inhibited M/TC synapses. These now permit transmission, initiating the hypothalamic output which disrupts pregnancy. The result of raising the level of activity at reciprocal synapses is to discriminate signals from individual males. The glomeruli which are activated by either-sex signals are sited in the anterior AOB region of the opposite sex. VN-x of females removes the sex-specific input confirming the restriction of the pathway for such signals (Dudley, 1999).

Inhibitory controls are found in the glomerular microcircuits which underlie the PB effect, and could represent a common model for the local functions of the AOB (Brennan, 1990; Taylor, 1991). Indeed, its simplicity has led to suggestions that it represents an evolutionary conserved neural subset (Fig. 5.4) for mate-recognition (Keverne, 1990 and 1996). Details of bulbar circuitry in other AOS-dependent mammals are needed to evaluate the generality of this model (Dudley, 1996).

The ability of female mice, and to some extent rats, to detect male signal variants such as those arising from the MHC loci, have been intensively studied with contrasting results (Brown and Eklund, 1994). The recognition of volatiles which could be associated with such fine discrimination of inter-individual signals may be processed jointly by each section of the AOB, and in the MOB. Eight volatile carboxylic acids occurred in significantly different relative concentrations across 12 separate mouse urine samples. Each of these came from males who differed only at an MHC locus. The results indicate that even at this level of genetic uniformity, pattern-discrimination is a probable basis for strain identification and or individuality (Singer *et al.*, 1997).

5.3.5 Populations

The inter-female Lee-Boot Effect is identified as the occurrence of acyclicity in groups of at least six (mice), if kept without male stimuli. Lone females are also susceptible to the chemosignals left in a cage previously occupied by "crowded" females. Estrous cycle suppression under these conditions is mediated by adrenal secretions voided in urine and assessed by VNO uptake. Mature female mice do not respond to urine from adrenalectomised females. Nasal application of urine from intact group-living females prevented cyclicity, as would their presence (Ma *et al.*, 1998). Out of the six active volatiles (Chap. 3) isolated from crowded female urine, one: 2,5-dimethylpyrazine (DMP), was alone effective as an estrous inhibitor. The presence of DMP would appear to signal a "no-male" condition, i.e. one unpropitious for immediate breeding. Conservation of reproductive effort, under even temporary adverse circumstances, is presumably of advantage since it avoids the risk of litter loss and impaired productivity. Under high densities, the urine from female wild mice contains the factors which delay female puberty; possibly an adaptive limitation on recruitment to the breeding stock (Massey, 1980; Coppola, 1985 and 1987).

Adrenal-mediated effects occur within the multimale groupings of Mouse-lemurs within which access to estrous females is confined to the dominant male (Perret, 1995). Subordinates are gonadally suppressed, and when exposed to attack show the expected adrenocortical arousal (Perret, 1992). The consequent urinary output contains the male–male inhibitory signals (Chap. 7.3).

Sociogenic effects are not confined to the eutherian (higher) mammals; they have been found in the, regrettably few, marsupial studies. Olfactory-reproductive effects appear in Grey Short-Tailed Opossums (*Monodelphis domestica*): exposure of isolated females to males induces female maturation and early estrous (Fadem, 1989; Stonerook and Harder, 1992). After VN-x, 70% remained pro- or anestrous, whereas incompletely VN-x females (n = 2) and control opossums responded within seven days of exposure to male scent (Jackson and Harder, 1996). The same procedure found that prolonged (~one week) direct exposure to a male could eventually override the vomeronasal-induced block (Pelengaris *et al.*, 1992). In *M. domestica*, proceptive behaviours, and surprisingly ovulation, were not impaired by VN-x — only by chemical MOE-x. Other opossums of this group display reproductive alterations with similarities to the rodent phenomenon. Potential olfactory-related changes occur in Woolly opossums (*Caluromys philander*), captive males show synchronicity of plasma T peaks, as well as an estrous-induced rise in testicular activity within pairs (Perret and M`Barek, 1991).

The variety of breeding patterns displayed by the American and Australian metatheria and the lack of comparative experimentation hampers general conclusions on any fundamental division from eutherian mammals.

In the context of house mice living in natural or semi-natural groupings, the role of male dominance is seen as crucial, in particular, when alpha males contribute most urinary stimuli through depositing a high density of marks (Hurst, 1989). The replacement of an "old" by a "new" *alpha* male during territorial encounters is one situation which favours operation of PB effects (Chipman *et al.*, 1966). The recipient female may be provided with a fitness assessment of the intruding male which is to the genetic advantage of subsequent litters (Hurst *et al.*, 1994). The urinary component for information transfer relating to male individuality, status, etc. is the (sex-specific) protein fraction

(Vandenbergh, 1975; Marchlewska-Koj, 1981; Mucignat-Caretta *et al.*, 1995).

5.3.6 Central Effects

Reproductive integration of volatile versus non-volatile inputs [Fig. 5.15(b)] is identified as those response abilities which can still be elicited after peripheral lesions. Support for differential integration comes from localised chemical (NMA) treatment of specific regions of the amygdala. Selective damage to the medial nucleus decreased paternal behaviour in a microtine rodent (Prairie vole). Neurones in the cortico-medial nucleus of the amygdala appear to be essential for maintaining "partner" female contact as well as responses to a pup, whereas the baso-lateral nucleus was not involved with intrafamilial behaviours (Kirkpatrick, 1994). Direct effects on sexual behaviour mediated by the medial amygdala (mAMYG) are evident in the receptive posture of female rats. In ovariectomised rats, effects on lordosis were evident from 10 to 50 days post-operation (Dudley, 1994).

The AOS in male mice transmits stimuli from soiled bedding to the PMC in the female amygdala, which does not respond with increased fos-ir, whereas the PM(v) does (Yokosuka, 1999). Anaesthetising the test female does allow direct contact by the VN-x males who then persist with preferences for a novel (strange) female, even after sexual satiety (Johnston, 1984). The appropriate cue here derives from the sebaceous (dimorphic) flank glands, rather than from reproductive tract sources. Male mice chemosignals in intact urine, and MUP without two of the male signals, induced responsiveness mainly in the rostral part of AOB, but partially in its posterior region (Brennan, 1999). The ligands alone (DHB and BDT), activated exclusively posterior responsiveness. The distribution of pathways carrying separate sex-signals supports the view that the female bulb in mice shows regionally distinct abilities in processing urinary signals. Brennan (1995) found that female mitral cells are capable of distinguishing the input arising from male signals by a sustained dis-inhibition; re-exposure to stimuli from the same male is potentiated over a two-day period, whereas following mating, the usual responsiveness noted in the Bruce effect is the enhanced inhibition mediated by GABA. The chemosignal effect of male urine involves a rise in the excitatory transmitters Glu and Asp at the reciprocal synapses between mitral/tufted cells [Fig. 5.12(b)].

Alterations in synaptic morphology could well provide the basis for the selective enhancement of sex-signal processing (Matsuoka *et al.*, 1998). As seen in female mice and rats and in male hamsters, the initial excitatory synapses at mitral/tufted cells could be potentiated by increases in size. Local structural expansion is presumably directly associated with the enhancement of transmitter production. The inhibitory symmetrical synapses did not show any such change following exposure to heterospecific urine. Confirmation of a functional linkage between chemosignal processing and adaptive enlargement would be required for other types of semiochemical.

5.3.7 Vomeronasal Independence

Several studies have identified responses that do not involve VN participation, from marsupials to Mouse-lemurs. Where the chosen end-point is totally unaffected by absence of the organ and in addition is dependent upon MOS activity, then it needs to be classified as VN-independent. Where VN-x results are ambiguous, as already considered for opossums (*Monodelphis domestica*), further analysis is desirable. For instance, Goats do not use AOS input for mating, only urinalysis, although experiential variables have not been fully explored (Ladewig *et al.*, 1980). Examples of VN independence then exist in both altricial and precocial species.

In situations where close-contact chemocommunication could reasonably be expected, as with infant–mother orientation, lack of AOS participation is all the more surprising. In rabbit pups, orientation to suckling is not interrupted by VN-x (Hudson and Distel, 1986). Responsiveness to amniotic fluid however, has a role in the maternal behaviour of primiparous/multiparous ewes, which persists after MOE-x (Levy and Poindron, 1987). In a recent study, the relative importance of AOS versus MOS for lamb recognition produced alternative findings. The VN-x technique was cauterisation of the nasoincisive duct (N-Pd) in pregnant ewes, this inhibited the normal rejection of alien lambs

[Fig. 7.10(c)]. These ewes allowed both alien lambs and their own lambs to suckle (Booth *et al.*, 2000). Since MOE-x still permitted own/alien discrimination, it appears that lamb acceptance is dependent upon the ability of the AOS to distinguish familiarity. The discrepancies may derive from technique or strain differences. The combined, possibly central, effects of complete deafferentation are perhaps necessary to disrupt such important responses. Whether, for instance, multiparous rats could be at all resistant to such treatments has not been tested.

Carnivores are as strongly odour-oriented as any other mammalian group (Gorman and Trowbridge, 1989; Weiler *et al.*, 1999). Detection of sex-related signals from soiled bedding in domestic polecats (*Mustela puto*) does not involve activation of the AOB. Estrous or male odours reach central nuclei, while en route activating those sites in the MOB where the granule cells possess an androgen receptor (Kelliher *et al.*, 1998). The absence of AOS processing in a reproductive context may point to its lack of androgen-dependent circuits in the ferret.

Anti-male attacks can be induced towards familiar (= recognisable) individuals. After VN-x, female mice though, did not attack unfamiliar males whether the operation was performed before or after mating, the effect lasting well into the lactational period (Bean and Wysocki, 1985, 1987 and 1989). Males suffer a similar diminution of aggression associated with loss of VN input (Clancy *et al.*, 1984). Again, although this effect was already known as a consequence of complete anosmia, the presence of a role for VNO-relayed chemosignals was unexpected.

A selective effect on litter defence by the suckling female is clearly of immediate survival value. Related behaviours such as aggression, also suffer interruption when the amygdala nuclei are destroyed. If performed before mating, there were no subsequent effects on rat maternal behaviour during the normal lactation period (Luiten *et al.*, 1985; Kolunie and Stern, 1995). Chemical lesions of the MOE were also ineffectual; only complete ablation during pregnancy was shown to be of any importance — decreasing aggression and affecting pup survival through loss of attentiveness by the lesioned mother rat.

A wider variety of rearing strategies across representatives of all diosmic taxa would extend assessments of the relative chemoreceptive contributions to parental–offspring interactions.

Some effects on inter-male aggression occur in Mouse-lemurs: after VN-x, attacks are significantly reduced in the presence of females, possibly as part of a more general inhibitory CNS effect (Aujard, 1997). Chemoinvestigation responses were not significantly reduced by VN-x, whereas direct female-oriented responses were affected. The frequency of the pre-intromission patterns (anogenital investigations and mounts) was lowered, but not sufficiently to impair successful mating. Presumably plasma testosterone levels were maintained in experienced (alpha) males, as in rodents.

5.3.8 Man

The abilities of human primates to make sexual or other discriminations are shown in a few, relatively weak, olfactory effects with minimal vomerolfactory content. The VN genome in man (Chap. 6) is so far the smallest known and, if active in adults, the lack of an AOB implies input via the MOB or an extra-bulbar tract (Humphrey, 1940; Meisami *et al.*, 1998). The "displacement" of a human VNOR gene to the main olfactory cell sheet (Chap. 6.3) provides a clue to the possible location of "lost" AOS functions. Any microvillar cells and the MOE "pit" provide likely candidate sites (Moran *et al.*, 1982; Feng *et al.*, 1997). Whether this is a solely human solution or common to the anthropoids and possibly other VN-diminished groups (Chap. 1), will no doubt soon emerge. The VNO in man has and will continue to be an enigmatic feature; estimates of its status range from entirely non-functional at all stages of development and into adulthood, to of limited pre-natal sensory capability, and to limited-response in some adults.

The balance of the evidence at present inclines against any major chemosensory role (Monti-Bloch *et al.*, 1998; Trotier *et al.*, 2000; Meredith, 2001). As noted, evidence of pre-natal, even if transient, functionality (Chap. 4) needs expansion not neglect (Yukimatsu *et al.*, 2000). Its existence into adult hood is at least anatomically admitted, while the degree of variability uncovered in recent surveys of occurrence and of basic morphology (Table 5.1), suggest that an absolute functional disregard is premature.

Sensory processing of semiochemical stimuli, by the accessory pathway, is achieved by amplifying and streamlining the passage of

Table 5.1 Occurrence of adult human vomeronasal structures.

N.	Age/sex	Incidence (%)				Investigations					Reference
		visual	endoscop.	laterality Bilat.	laterality Unil.	Techn.	VNd	VNgl	VNE	Others	
200	n/a	16 [inspect.]	60	91	9	TEM [N = 1]	+	+	L & D cells	*-ve* cilia	Gaafar *et al.* (1998)
14	n.s.	100 [TEM]				TEM		+ (serous)	L. cells + neuro-filaments + mv	basal axons	Jahnke and Merker (1998)
173	2-91/ n.s.		60	39	21	MRI [N = 11]	+ (3/11)		lumen patent (8/11)	L<->R inter-lumen (2/8)*	Hummel *et al.* (1999)
100	n/a		28.2			Histol.					Won *et al.* (2000)
1842	ns		39	13	26	Imm. histoch.		+	*-ve* OMP & S100	Keratin *+ve*	Trotier *et al.* (2000)
253	ns	6	?	36	64				non-senescent		Zbar *et al.* (2000)
12	n.s.	91.6 (LM)				Histol.	multiple	PAS *+ve*	vol. = 0.5 VNO	*-ve* mv; *+ve* cilia	Smith *et al.* (1998)

*Bilateral connectivity of VN organs. L & D = Light and dark cells.
n.s. = Non-significant; n/a = not available; +ve = present; – ve = absent/not recorded.

information to restricted areas of the brain. Although structurally and functionally distinct, both systems are equally capable of handling the same types of information and of modulation of the Hypothalamic-Pituitary-Gonadal axis.

Integration of exocrine and endocrine stimuli is not exclusively the property of either system. The extent of their mutual contribution(s) is unresolved; some stimuli will be highly specific to one of the systems, some will activate both — the relative proportion of activation varying widely.

6 MOLECULAR BIOLOGY

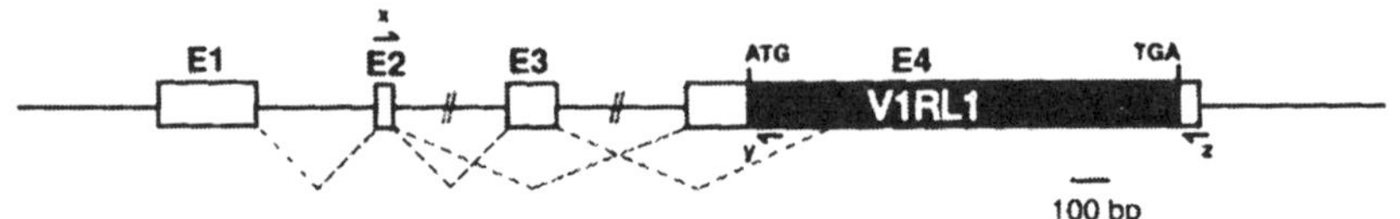

Coding region of the vomeronasal (V1R) gene, as expressed in human MOE neurones (from Rodriguez *et al.*, 2000).

"A nose that can see is worth two that can sniff".

Eugène Ionescu (1965)

6.1 INTRODUCTION

This chapter deals with the genetic requirements of accessory chemoreception, its involvement in the functional division of the AOS and the implications of its divergence from the MOS. The organisation of the mammalian sub-genome may well reflect the selective changes undergone as vertebrate complexity and adaptiveness increased. Olfactory plus vomeronasal alleles take up a major part of the genome. The chromosomal regions which code for chemoreception rival the size of the immune gene array (about 1% of the total), a comparable system which also deals with extraneous and random chemical bombardment. The size of the MOE gene bank reflects the range of compounds impinging on it (Chap. 3). In the AOS, evolutionary changes have involved a gradual segregation of function, and in consequence, the two chemoreceptive sub-genomes exist with apparently little overlap in their receptor repertoire. There is no clear indication as to the likely origins of vomodorous molecules, and of the genetic control which allows their detection. A few pointers can be derived from the make-up of olfactory receptors (ORs) found in the primitive jawless fish, the lobe-finned Coelacanths and other "living fossils", such as the Lung-fish (Berghard and Dryer, 1998; Freitag, 1998). Distinct VN-like neurones (Chap. 5) already appear in segregated areas in some, but not all, advanced (bony) fishes (Cao *et al.*, 1998; Asano-Miyoshi *et al.*, 2000). Molecular evolution of receptors seems to have preceded morphological specialisation.

The anatomical separation achieved by reptiles parallels the divergence of the main/accessory receptor categories. AOS chemoreception is assumed to elaborate its semiochemical responses by selective gains in ligand–capture efficiency and by alterations of threshold values. Once

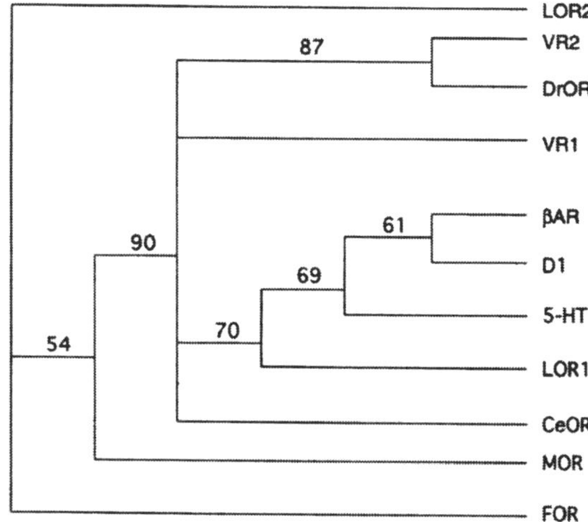

Fig. 6.1 Interrelationships of chemoreceptors: internal (neurotransmitters) and external chemosignals. Phylogenetic connections for sequences in transmembrane (Fig. 6.2) domains; Nos. = bootstrap values from 100 Megaline searches (based on majority consensus tree). Invertebrate — DrOR fruit-fly, CeOR nematode; vertebrate — FOR fish, LOR (1 & 2) lamprey, MOR mouse; VR (1 & 2) vomeronasal (from Dryer and Berghard, 1999).

Table 6.1 Organisation of accessory chemoreception.

VN cell body zone	Projection site of VN axons	Gene category	Genome size (%)	Chemosignal molecules	GPCR variant	Ion channel
Upper (apical)	Anterior AOB (rostral)	V1R	30–35 (25)	Volatile (hydrophilic)	Gα(i2)	sTRP.2
Lower (basal)	Posterior AOB (caudal)	V2R	130–150 (75)	Non-volatile (hydrophobic)	Gα(o)	sTRP.2

Note: Organisation is based on the AOS of rodents [cf. Fig. 5.15(a)].

established, the OR complement of alleles seems resistant to selective action. Despite deliberate selection for olfactory abilities, there is little induced variation apparent in the dog genome. Across 26 domestic breeds, the number of genes in each of the four subfamilies — two up to 20 members per subfamily — were stable in OR numbers (Issel-Tarver, 1996). In contrast, the influence of selection on VNR composition might be sought among groups with distinctive reproductive strategies. The emergence of vomeronasal input as a major regulator over key reproductive neurocrine systems clearly relates to the importance of its contribution to species survival (Keverne, 1999).

6.2 STRUCTURES

Olfactory and vomeronasal receptor cells have heptahelical trans-membrane (7-TM) receptors which belong to the largest group of G-Protein-Coupled Receptors (GPCRs) (Mombaerts, 1999). These provide the initial step in the transduction mechanism which allows chemosensors to make specific responses to extracellular messages (Josefsson, 1999; Dryer, 1999). The secondary structures encountered in the 7-TM domains are highly variable; the basic OR arrangement [Fig. 6.2(b)] has an external NH-terminal segment with interconnecting loops. The nearest comparable structure on other excitable membrane sites would seem to be the calcium-sensing and metabotropic glutamate receptors. The VN gene families are distinct from the MOE alleles and fall into two categories. One of the G-proteins is constructed with a large (550 amino

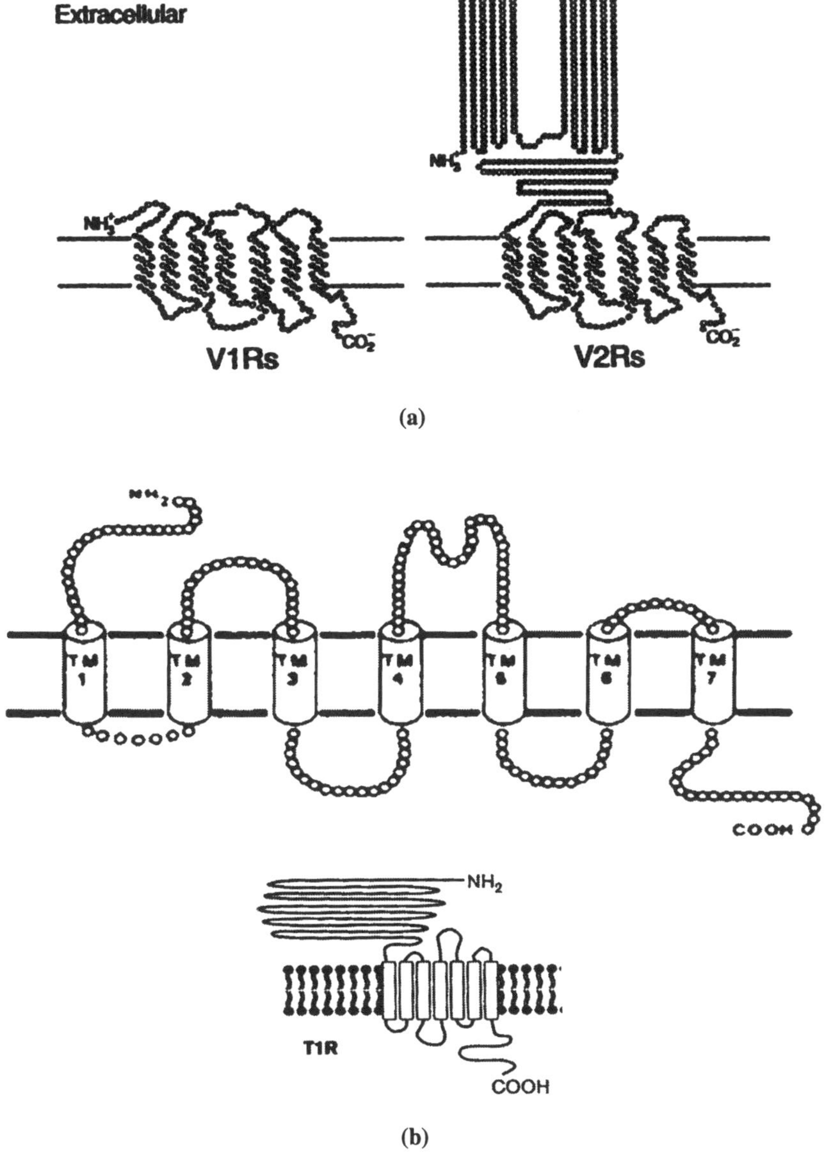

Fig. 6.2 Topology of GPCRs for the main classes of chemoreceptors. Group A: vomeronasal (V1R and V2R) and Group B: olfactory (OR) and taste (T1R) (from Tirindelli, 1998; Hoon, 1999; Gilbertson, 2000).

acids) external polypeptide V2R [Fig. 6.2(a)]; this domain has few similarities with any other GPCRs (Ryba and Tirindelli, 1997). The primary variability and secondary structure of the receptor's external "tail" and of a near transmembrane domain (no. 3) are very probably designed to fit the requirements of its ligand-binding capacity. The other receptor, VN1R, resembles the main OR-GPCR in tertiary structure only [Figs. 6.2(a) and (b)]. Few convincing models have been tested on VN-specific molecules for their binding requirements to receptor sites (c.f. Afshar, 1998). The location of the extracellular binding site is defined by the arrangement of side-chains on segments 3 to 5 of the GPCR, the probable nesting position of a hypothetical semiochemical is shown in Fig. 6.3 (c.f. Pl. 3.1B).

The vomeronasal receptors are assigned to a small sub-group of GPCRs, and contain two types of α-subunits, each being expressed in a distinct layer within the neuroepithelium (Dulac and Axel, 1995; Matsunami and Buck, 1997; Bargmann, 1997). The cells nearest the basal layer form a band which expresses the Gαo (Go) genes, while an upper (apical) layer of neurones shows an alternate version — the Gi2 (Gi). The positional expression of each GCPR type was established by *in situ* hybridisation probes (ISH) and immunohistochemistry (Halpern, 1995; Berghard and Buck, 1996). In the upper neurones, the VNOR genes are labelled as V1R and in the lower zone as V2R. As expected, there is co-localisation in the neuroepithelium of the corresponding

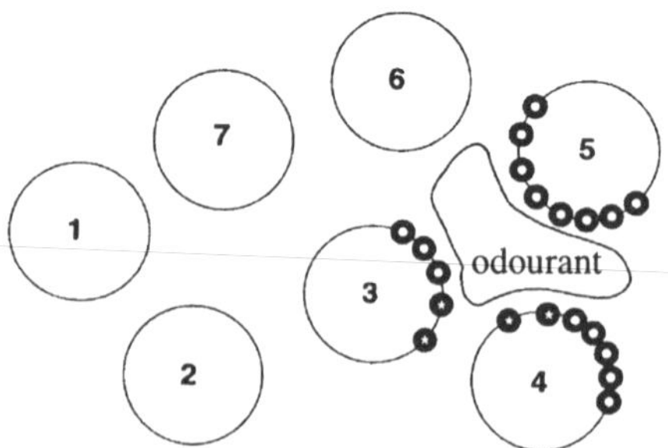

Fig. 6.3 7-Transmembrane GPCR and extra-cellular view of odourant molecule: shown in putative binding position: hydrophobic and hydrophilic side-chains (**O**) shown for protein segments 3, 4 and 5 (from Sharon, 1998).

mRNAs for VN1 and VN2, identified by ISH, and derived from the 3' non-coding regions (Saito *et al.*, 1998). The neuronal layers differ in gross morphology since the lower cells extend long dendritic processes, as opposed to upper cells. Both types of neurones express odourant receptor molecules on their microvilli; some, such as the rat VN6 peptide, can also be found on dendritic knobs (Menco, 1997; Takigami, 1999).

A likely ion-channel, the short TRP-2 version (Chap. 5, Heading Fig.), is now directly linked to the VN transduction process and does locate specifically to the microvilli (Liman, 1999; Harteneck *et al.*, 2000).

6.3 REGIONAL EXPRESSION

The GPCR pattern in the VNO shows that some overlap in the distribution of the two receptor types occurs. The degree of mixing does not reach the extent of the random assortment amongst the groupings of neurones found in the MOE. This segregation contrasts somewhat with the overlapping expression of genes in the main olfactory system (Buck, 2000). The Gi locus is restricted to olfactory neurones along the dorsal septum and recess, whereas Go is universally expressed across the whole MOE (Wekesa and Anholt, 1999). Amongst the main primary neurones, segregation of the 1000 to 1500 Golf genes occurs within rostral-to-caudal bands (Ressler *et al.*, 1993). These OR zones comprise at least four sets or clusters with about 25% of the OR complement being restricted to each zone, i.e. any one gene is poorly or not expressed in any of the other three zones. However further analysis, for instance by the examination of OR gene chromosomal sites, suggests that occasional clustered loci exist. A linked gene group being reflected in the OR cell localisation pattern as small MOE foci (Strotmann *et al.*, 1999). The OR multigenes in the rat have coding regions which are continuous since there is an absence of introns; only the 5' untranslated regions possess introns (Buck, 1993). A single allele and its receptor are thus scattered randomly across each receptor cell sheet. This type of distribution may maximise the pick-up of all classes of odourants (c.f. Chap. 3 {**1.–12.**}). The expression pattern found in mice and rats is not yet confirmed for other species (Masumoto *et al.*, 1999). Clearly, a range of chemoreceptive epithelial organisation should reveal some of the selective advantages involved. Heterogeneity within the AOS of primitive mammals, one of the opossums, suggests that segregation amongst both the primary and secondary neurones was an early establishment in mammalian organisation (Halpern *et al.*, 1995 and 1998). The rodent AOS also demonstrates the persistence of partitioning within the neuronal groups of the organ and of the AOB (Table 5.1).

In one ungulate (goat) AOS, the distribution of the (Gi) group of the GPCRs axons is uniform, not segregated, in the accessory bulb (Takigami *et al.*, 2000). The other class of primary VN neurones was absent since Go-immunoreactivity was suprisingly not identified in either the VNO or the VN nerve layer of the AOB. This lack of the heterogeneity, which typifies domestic rodents, may represent just one of many mammalian variants. In adult rodents, the diagnostic chemoreceptive cell marker (OMP) also shows a uniform gradient of OMP+ve cells: from highest rostrally to lowest caudally; again without any sharp demarcation boundary within the bulb. An additional distinction between MOE and VNE expression occurs with the G protein γ-subunit/8. In rats, this gene occurs in adult VN neurones, but only in immature ORs; a developmental divergence of unknown importance (Ryba and Tirindelli, 1996). Maturational distinctions within the VN gene groups relate to male/female differentiation, again in Go(V2) localisation (Herrada and Dulac, 1997). Females of the domestic rat show a larger expression band than males, and one which is closer to the Gi zone. Further work could identify whether this sexual distinction is due to a sex-specific sub-population or to the presence of gene variation related to the discrimination of chemosignals distinguishing the sexes.

The heterogeneity of the VN primary neurones is reflected in their modes of chemosensory preferences. The relative binding efficiencies for distinct odourant types onto the membrane sites is indeed functionally partitioned. When urinary fractions from male mice were applied to VN cells of females, stimulation by a lipophilic and volatile odourant fraction activated only the Gi protein-expressing cells. In contrast, Go activation was elicited by one of the lipocalin superfamily; the MUP fraction containing an α-2-globulin (Krieger, 1999). This observation partially illuminates the ability of short segments of MUP to activate some VN receptor sites concerned with neuroendocrine modulation (Mucignat-Caretta *et al.*, 1995). It would be intriguing to discover whether there are specific carrier molecules (Chap. 3) in the surface mucous layer which are associated with the recognition of particular odourant-types. Evidence of this kind is provided by the occurrence of some sexually differentiated MUPs, produced in mouse liver, and recently identified as being among the nasal versions of OBPs (Held *et al.*, 1987; Ohno *et al.*, 1996). Such findings also support the hypothesis of a functional division amongst the VN bands. Each band may have the propensity to respond preferentially to a distinct vomeromodulin–ligand complex. In the VN2 cells, the receptors are tuned to the more typical non-volatile VN odours as "large molecule" specialists. It is likely that the VN1 receptors, which resemble the main ORs in structure are also functionally similar in responding to smaller volatile odour molecules (Krieger *et al.*, 1999). Within each VN broad band, there are subsets of neurones arranged in several "distinct hemiconcentric zones of different radii" (Herrada and Dulac, 1997). To explain the intra-epithelial assortment of function, it will be necessary to provide some association of functionally discrete roles with each of the internal VN units containing Go/Gi variants. The application of AOS-specific semiochemicals to isolated VN slices (Chap. 5) points to the solution of this intriguing question. Six pure volatile compounds from the urine of male and female mice activated discrete sets of VN neurones, all of which were located in the Gi band (Leinders-Zufall *et al.*, 2000). It now seems that there is indeed a regional specialisation of response for many rodent chemosignals. Are the molecular species recognised by each of the zones related to one another either in structure or in function — or both?

6.4 HUMAN VN GENES

A surprising and unique feature of the human VN genome is not its small size, but that it exists at all. Of the eight loci located, seven are inoperative, due to frame-shifts and/or stop codons; the one potentially functional gene, named V1RL1, was found in all individuals from a random sample (N = 11). Most significantly for a general analysis of the VN sub-genome, this single remaining gene was expressed within the MOE, not in any VN cells (Rodriguez *et al.*, 2000; see also "Note Added in Proof" section, pp. 148). This further suggests a residual, presumably non-pheromonal, role for any chemoresponsive neural tissue remaining in the adult human VNO (Trotier *et al.*, 2000). The identification of an apparently dislocated remnant of the VN genome in the adult should be set against the finding of positive foetal Go and Gi expression at mid-gestation (Yukimatsu *et al.*, 2000). A hypothesis for this apparent anomaly lies with the events which occur during this critical period in development. The persistence of VN and *N. terminalis* axons is still required, as they are acting as "route-markers" for the migratory stream of GnRH neurocrine cells (Chap. 4.5), seemingly only during uterine life. The loss of active VN genes in the tissues of the adult organ may result from pre-programmed, and at least prepubertal, neuronal death; following regression of the adult AOB (Meisami *et al.*, 1998). There is in consequence a loss of the secondary bulbar neurones as targets. This

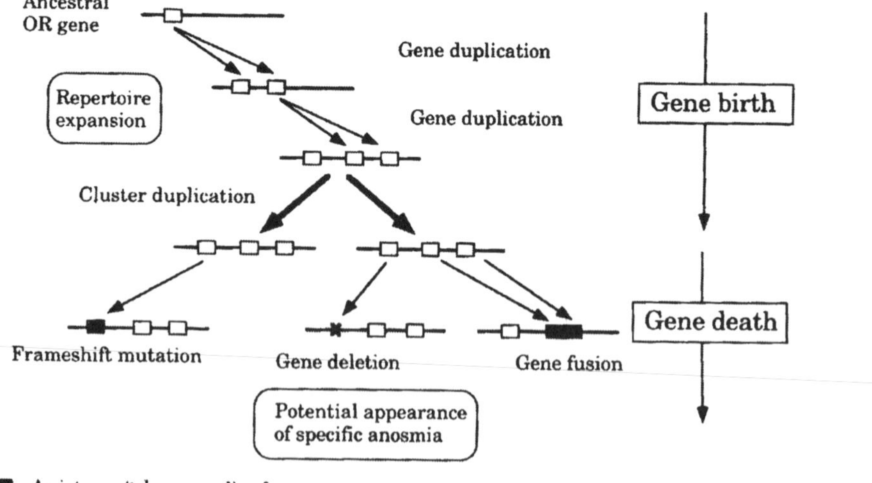

Fig. 6.4 Evolutionary schema for the emergence of the OR genome: increase, elimination and alterations to the main receptor repertoire (from Sharon, 1998).

deficit, plus the lack of replacement mitotic activity in the basal layer provides some explanation of the developmental decline in VN chemoreception. The accumulation of non-functional and defective VN alleles may be the consequence of the reversal of a trend toward extension of the overall OR complement by gene duplication and variation (Ben-Arie *et al.*, 1994). The phylogenetic origins of these changes can be set in the period of divergence of the New- from the Old-World monkeys, and of their hominoid descendants (Fig. 1.4). Once full comparisons are possible with the other higher primates, it should be possible to examine the intermediate steps in the downgrading of chemoreception in man and apes to microsmatic status.

The decline has involved the inclusion of about ≈40% pseudogenes in the OR repertoire (Fig. 6.4); many having undergone gene conversion during the reduction process (Sharon *et al.*, 1999). Chimpanzee alleles resembling human sequences locate to chromosome 19/p15; within the OR clusters considerable segment shuffling has occurred. This process in principle resembles the degree of diversification in gene combinations reached by immune loci.

Further longitudinal studies of VNOR developmental changes, and of the early action of genetic defects, are urgently required to define the period of receptor switch-off in higher primates. Until the extent of prenatal AOS contributions to ontogenic sequences can be clarified in other mammals, then the operation of the remnant human VN function will remain obscure. As noted earlier (Sec. 5.4.1), responses to the proposed semiochemical steroids in man need to be re-evaluated in view of the location of a functional allele without the organ itself. Should this permit signal detection, then at least some pheromonal phenomena, such as menstrual synchrony, will approach a satisfactory explanation.

6.5 CHROMOSOMAL DISTRIBUTION

The location is now known for an increasing proportion of the OR genome (Ben-Arie *et al.*, 1994; Strotmann *et al.*, 1999). Most of the human olfactory receptor genes are located on chromosomes 6, 11 and 17, and inhabit sites on all others except 18, 20, X/Y (Buettner *et al.*,

1998). At least 25 loci occur on Chromosome 11; these are found clustered in as few as seven distinct regions. The significance of the location of a batch of ORs will become clearer as mapping progresses. In particular, one or more sequences of an OR gene occur within the MHC-1 complex (Fan *et al.*, 1995). The class-I MHC locus in mice and men is provisionally attributed a role with the sensory evaluation embedded in mate selection or similar functions (Manning *et al.*, 1992; Grob *et al.*, 1998). This association may be fortuitous, but a functional link should be sought (Singh *et al.*, 1988; Brown and Eklund, 1994). The shrunken human complement of approximately 200–300 OR sites is thought to incorporate a majority of mutant alleles as pseudogenes (Crowe *et al.*, 1996; Mombaerts, 1999). The status and implications of findings on the unusually high incidence (estimated range 30% to 50%) of OR pseudogenes in *H. sapiens*, are discussed by Rouquier *et al.* (1998). To reveal details of the developmental organisation, underlying events such as the partition of the AOS will require mining of the genome map for non-human primates (Glusman *et al.*, 2000). In view of the diminution in man to a single VNR, and more particularly its displacement to the MOE, the accessory and main receptors should be fully compared (c.f. "Note Added in Proof"). Similarly, valuable evolutionary sequences will arise from analyses of intra- and inter-specific variation in VN sequences within highly variant taxa such as bats and non-human primates.

6.6 CELLULAR EXPRESSION

The important regulatory elements of OR genes are located quite near to their transcription initiation sites and may mediate and receive information defining the laying down of zonal patterning, possibly via long-range effects (Qasba and Reed, 1998). This model cannot yet be applied to AOS morphogenesis until a complete VN allele map has been filled out. Insertion of a control transgene in mice suggests that long-range gene regulation exists and that it can direct OR and VNOR expression, being itself also subject to allelic inactivation (Ebrahimi *et al.*, 2000).

The mechanism which operates to select the functional member of an allele pair during neuronal maturation is not known for vomeronasal

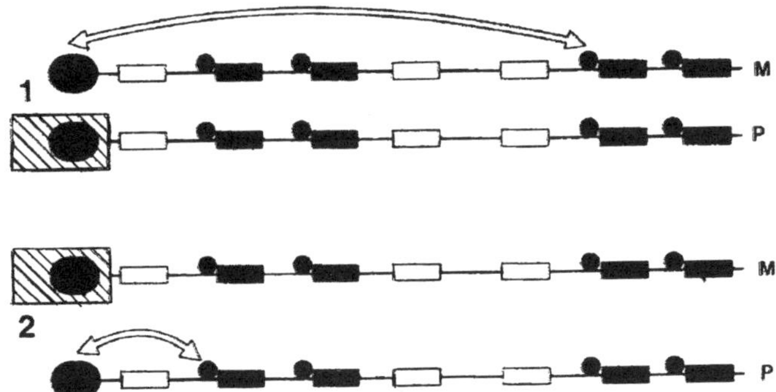

Fig. 6.5(a) OR gene expression. M, maternal and P, paternal DNA strands; 1 & 2, chromosomes from neurones with differential *cis/trans* regulation (M/P) of each DNA strand (from Chess *et al.*, 1998). [■ = OR gene locus; ▨ = suppressed regulatory element]

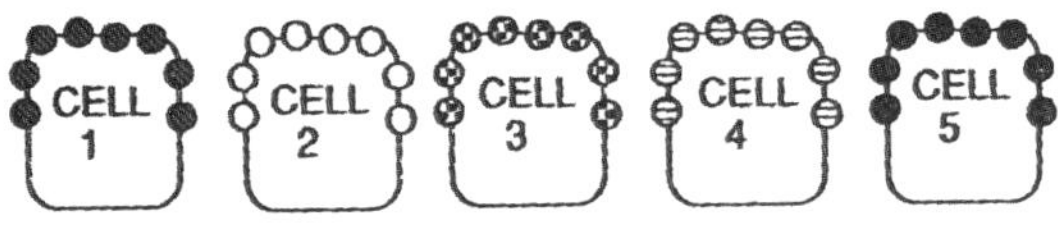

Fig. 6.5(b) Clonal exclusion (five OR cells): selective expression of one gene per cell out of five alleles (from Lancet *et al.*, 1993).

genes, but it presumably does not deviate substantially from that in olfactory neurones (Chess *et al.*, 1994). Each allele has a probability of expression of 0.5, since only one member at each locus is active, there being random suppression of the partner allele. A similar process, random X-chromosome inactivation, takes place in the tissues of mammalian females. Hence, any one OR neurone is monoallelic once the locus is expressed, and already occurs in that form in the germ-line, as confirmed in mice (Rodriguez *et al.*, 1999). The VN genome's size is about one-fifth of the current lower estimate of ~1000 for the OR genes. In diosmic species, there could be up to 200 loci divided into about 25% VN1 and 75% VN2 loci (Buck, 2000; Dulac, 2000). The range which is available to the upper neurones is out of ~one-quarter (35–50 V1R alleles), while the lower neurones "choose" from three-quarters (~150 V2R alleles). A random selection mechanism [Figs. 6.5(a) and (b)] is presumed to operate during neurogenesis (Chess *et al.*, 1998). This restriction of activation ensures that each presumptive neurone contains (1) only one pair out of its VN family, and then (2) only one member of that pair is activated. The operation of clonal selection probably accounts for some semiochemical anosmias (c.f. Wysocki *et al.*, 1999).

It could be surmised that such an expression system is tolerated in the MOS, where the large receptor repertoire discriminates at least 10^5 odourants. A volatile semiochemical is expected to require about 10 MOE receptor types, each of varying sensitivity, to induce a response. Even assuming a lower ratio of odours to receptors implies that the VN genome can cope with far fewer chemosignal molecules. Genome size then could be a selective liability in the AOS (but see "Note Added in Proof"). Even though pseudogene sequences are rare, 50% of the neurones in the VNE will be inoperative for any one locus. However, AOS requirements may have been tailored by strict limitations on its physiological and behavioural roles. There is no accurate information on the number of VNORs required for the discrimination of a single vomodour. Most, if not all, chemosignals are translated centrally by the decoding of combinatorial messages (Malnic *et al.*, 1999). An estimated repertoire size of about 2000 "pheromone-like" chemicals still allows for substantial redundancy in the minimum receptor groupings required for signal detection and subsequent processing.

NOTE ADDED IN PROOF

Recent advances in chemosensory genome mapping have established that the accessory repertoire has now a revised upper estimate in the size of the sub-genome expressed in the Gαi2 neurones (Table 6.1). Rodriguez *et al.* (2001), report that the number of VN1R families recognised has increased from 4 to 12. These novel members of the mouse AOS complement comprise some 104 new genes, grouped into eight families and all expressed in VN neurones. The additional families are described as "...*extremely isolated*", and considerably extend the known sequence diversity within the V1R superfamily of the mouse genome. The functional import of these findings is proposed as reflecting the specialisation of receptor categories by (variously, Fig. 6.4) selective removal, and/or expansion events (Lane *et al.*, 2002); both processes differentially impinging on allele survival. The range of upper receptor genes now appears comparable with that found within the lower zone Go loci, assignment of the details of functional differentiation (Martini *et al.*, 2001) are anticipated.

7 *BEHAVIOUR*

Urine-washing in Lesser Mouse Lemur (*Microcebus murinus*)
(courtesy Helga Schulze©).

"The Guanacos have one singular habit, which is to me quite inexplicable; namely, that on successive days they drop their dung in the same defined heap".

Charles Darwin (1839)

7.1 CHEMOINVESTIGATION AND STIMULUS UPTAKE

7.1.1 Fish

An intriguing analogy has been made between a behavioural element in the searching behaviour of flounders (Pleuronectids) termed "coughing", and sniffing in mammals (Nevitt, 1991). For air-borne odours, sniff-sampling is assumed to concentrate patchily distributed odourants by the introduction of pulses at the mucus-ciliated interface. Similarly, flounder coughing occurs preferentially in the presence of food solutes in a filtered water stream, as does that of Gobies during courtship sequences (Tavolga, 1956). The mixed nature of the neuroepithelium in fishes makes it unlikely that water samples are then specifically directed towards the scattered functionally differentiated areas. Nevertheless, inter-sexual orientation to and discrimination of socially relevant chemicals are now recognised as highly sensitive responses (Sorensen, 1996). Further, disruption of social patterns related to schooling do follow MOE-x (Bardach and Todd, 1970). There may be discrete searching behaviours which facilitate uptake of particular classes of semiochemicals, but chemoreceptive mechanisms in fish (c.f. Chap. 2) are not yet clearly understood.

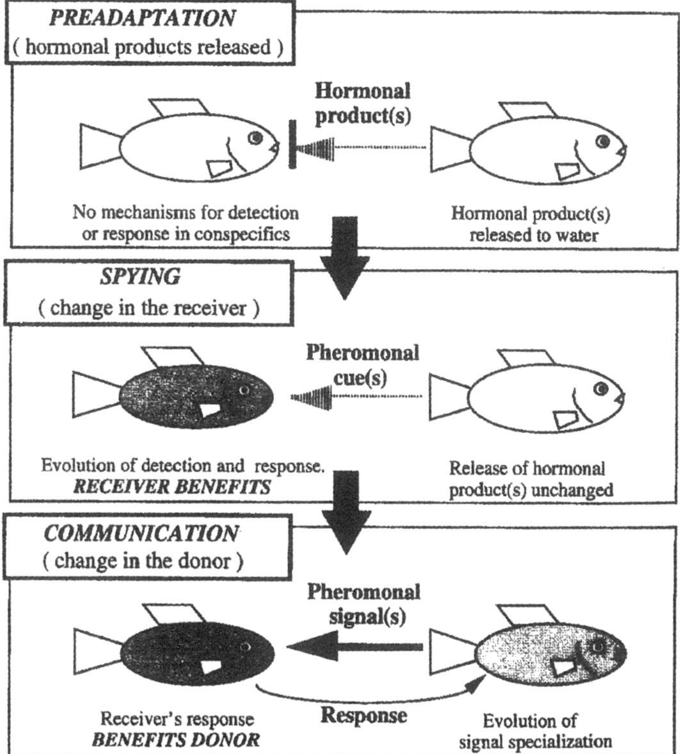

Fig. 7.1 Schema for emergence of signaller ≈ receiver transmission, selection for information exchange in, e.g. fish (from Sorenson and Stacey, 1999).

7.1.2 Amphibia

The newts, salamanders and caecilians, being the least terrestrial groups, are prominent and fish-like chemosignallers. Olfactory and/or vomerolfactory usage is most evident in those with wholly aquatic, or semi-aquatic habitats, and also in those where vision is restricted or inoperative. In contrast, Madison (1977) suggests that semiochemicals are least important in the auditory dominated anurans, although several, possibly exceptional, examples do occur (Chap. 3; and Wabnitz *et al.*, 1999; Pearl *et al.*, 2000). Larval stages have retained some of the anti-predator "fright" responses seen in some fish shoals. Tadpoles of the Leopard frog (*Rana utricularia*) show similar avoidance and dispersal patterns in response to chemical cues from predators (Lefcort, 1996). The abilities of breeding larvae (neotenous tadpoles, e.g. axolotls) have not been well studied for chemosignalling. In these forms, the presence of a clearly demarcated accessory (VN) chamber has been demonstrated (Eisthen, 1994). Hence, some partitioning of accessory responses could be expected in such wholly aquatic animals.

One well-analysed chemosignal system is that of the Red-bellied Newt (*Cynops pyrrhogaster*). Males of this species produce from the abdominal gland a semiochemical protein (sodefrin) with marked VNO activity as a female attractant (Kikuyama *et al.*, 1997). Courtship displays of newts often contain tail-waving bouts, which direct cloacal or other secretions from the male toward female recipients (Fig. 3.1). A large stable molecule like sodefrin, alone or as part of a VNPr complex, is presumably suitable for such local transference.

Signal conveyance by direct contact appears typical of aquatic forms, and is employed by several salamander species (Arnold, 1977). Removal of the female's epidermis by the action of the male's (dimorphic) scraper allows him to transfer mental (chin) gland secretions [Fig. 7.2(a)]. The effect of non-olfactory uptake is presumably immediate, and one of the very few internal signal paths known. A further example of direct transfer signalling is shown by Jordan's salamander (*P. jordanii*), where males smack the chin gland across the females' nostrils, presumably forcing secretion into the nares on contact. Amplexus (male clasping) is induced by a non-protein attractant in male Roughskin newts (*Taricha granola*), when presented with a model treated with female scent

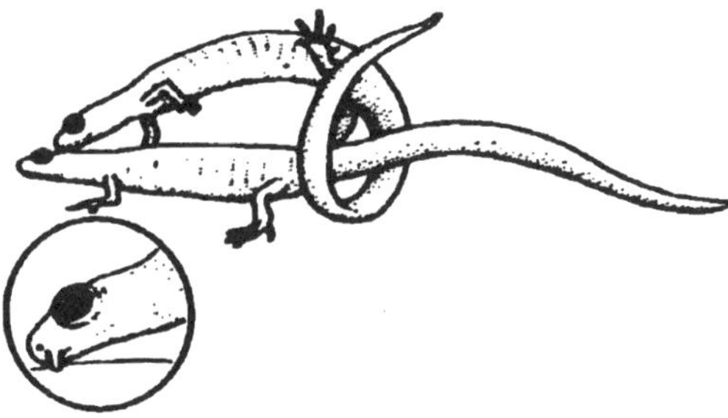

Fig. 7.2(a) Inter-sexual, direct scent transfer: signal-"injection", by scarification of female's skin. Male salamander transfers chin-gland secretion (inset) with dimorphic tooth-scraper (from Arnold, 1977).

(Thompson, 1999). A signal-gathering device for non-volatiles, the paired naso-labial grooves, exists in Plethodontid salamanders; sampled fluids travel along a hairline furrow extending from the edge of the upper lip to the boundary of the ipselateral nostril (Brown, 1968). These grooves then channel fluid into the nasal cavity where it accumulates in the accessory (VN) area (Dawley, 1989). Within the VN pocket (Fig. 2.7), ciliary action may be responsible for fluid movement across the receptor sheet (Eisthen, 1994). During chemoinvestigation, the male collects the female's chemosignals by direct uptake; either by head-tapping on the substrate or on the female herself.

An apparent convergence of lower- and higher-vertebrate postures is evident in the resemblance of signal transfer patterns which are functionally dissimilar. The partner-marking [Fig. 7.2(b)] of an arboreal prosimian, the Fork-marked lemur (*Phaner furcifer*) superficially mimics the Salamander pattern shown. The female is also a recipient of a directly transferred signal from the male, but here is merely a passive vehicle for a scent-broadcast. Transfer across furred skin is unlikely and the site of deposition avoids ingestion by licking.

Temporal alterations in peripheral chemoreceptors are rare in vertebrates, but this group provides an example of transient enhancement of signal capture efficiency. The Red-backed salamander (*Plethodon cinereus*) shows dimorphic and seasonal VNO volume fluctuations. Males always possess a significantly larger vomeronasal area, and both

Fig. 7.2(b) Inter-sexual, direct scent transfer: allomarking male → female in the Fork-Crowned Dwarf Lemur (*Phaner furcifer*), from Schilling (1990).

males and females show a summer expansion of vomeronasal volume (Dawley, 1995). The consequences of a timed increase in surface area could rest upon a pre-reproductive rise in GnTH which in turn probably facilitates increased receptor neurogenesis. Males and females presumably benefit from a suitably timed improvement in sensitivity to reproductive pheromones.

AOS responsiveness to hormonal influences is shown in the action of sodefrin on the lateral nasal sinus of newts (*Cynops*). The receptors in the accessory pocket are differentially affected by pituitary and ovarian hormones (Toyoda *et al.*, 2000). The local EOG response to the pheromone (Fig. 5.1) was enhanced by the presence of prolactin or of estrogen alone. Receptor sensitivity increase is perhaps an alternate strategy to AOS receptor density increase; several alternate routes of signal ↔ receptor adaptation (Fig. 7.1) have been hypothesised (Sorenson and Stacey, 1998).

The chemoreceptive mechanisms in amphibia are undoubtedly worthy of further analysis, not only for their own sake, but to provide clues as to the origination of advanced chemosignal systems. As noted above, a pheromonal signal from the mental gland acts as a courtship/receptivity inducer. The plethodontid receptivity factor (PRF); (Chap. 3) despite its size (22 kD), seems to have been "converted" from its internal role as an inter-cellular cytokine, to an inter-individual co-ordinator of reproductive activity (Rollmann *et al.*, 1999). Endocrine or other internal message-molecules are then likely candidates for inter-individual (exocrine) coordination, as in the steroidal metabolites released by teleosts. Amphibians may have adapted the biological activity of the defence peptides secreted by the skin.

7.1.3 Reptiles

The vomeronasal component of olfaction is very probably highest in this group and has long been implicated in chemoinvestigation (Wilde, 1938). A review of the early literature is given by Burghardt (1980), and of recent chemosensory biology by Halpern (1992) and Mason (1992). There is considerable diversity in their semiochemical usage, from the largely terrestrial and cryptic snakes, to the habitat generalists amongst lizards. Behavioural contributions extend from reproductive patterns to prey-location and anti-predator defences (Graves and Halpern, 1990). General social behaviour is spectacularly aided by conspecific recognition during mass hibernation events, when aggregations of up to 10^6 garter snakes occur. Taxon-specific discrimination trials showed that preferential responses were given to VN-detected skin lipids from conspecifics, but not to those from sympatrics such as Corn snakes (Graves *et al.*, 1991). As noted in previous chapters, evidence from interruption of stimulus access indicates a strong dependence on the AOS for a range of functions, not all being pheromonal in character. In

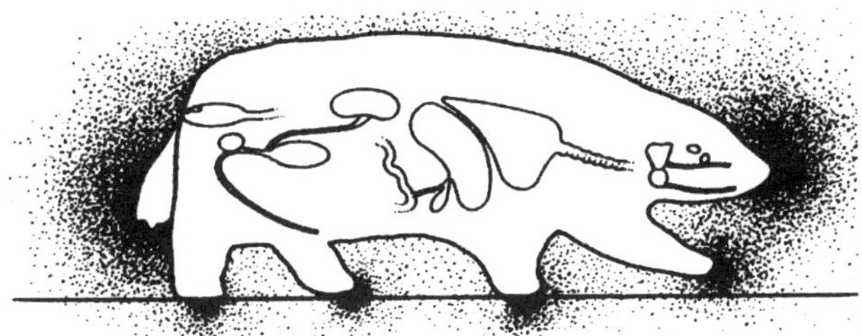

Fig. 7.2(c) "Vertebrate" signals, production and emission sites: generalised internal and external sources common to most terrestrial species, from oral (lungs, mouth and salivary glands) facial, Harderian glands; through liver and gall-bladder to alimentary and genito-urinary [o–>] tracts; skin surface including feet, specialised apocrine/eccrine scent glands not shown (after Flood, 1985).

the more primitive of the diosmic reptiles, the simple VNO of semi-aquatic turtles contributes largely to chemoreception in water. This finding was predicted from the pioneering recordings (Chap. 5) of non-olfactory activity in box-turtles, and recently confirmed (Tucker, 1963 and 1971; Hatanaka, 1993). The loss of a VNO-nasal connection in ophidians is thought to be related to the establishment of a forked tongue-tip (Schwenk, 1993). Indeed, vomerolfactory behaviour appears almost confined to tongue use, with chemoinvestigation virtually dependent upon tongue-flicking (TF). Considerable attention has been paid to the mechanics and the stimulus dependency of TF and also to its non-olfactory roles (Halpern, 1983; Graves, 1990; Greenberg, 1993; Cooper, 1994).

The operational variants in tongue movement so far detected, are a simple extension/retraction and oscillations — occurring singly or as multiple bouts Fig. 7.3(a); the latter type can be divided into those with slow motion and those of normal duration (Toubeau, 1994).

A complex of linked variables is associated with the elaboration of TF, the main features being the depth of tongue-tip division (its forkedness), relative tine and tongue lengths, plus the circular muscle system which controls the extent of its protrusion (Cooper, 1997b). Functionally, the tips of the fork play the most significant part in delivery of particulates [Fig. 7.3(b)] to the VN duct (Clarke, 1981).

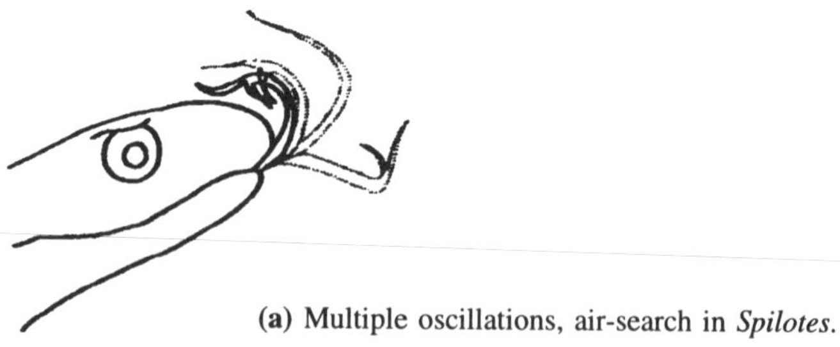

(a) Multiple oscillations, air-search in *Spilotes*.

(b) Sample contact, tine-tip uptake ● on lower surface) in *Tropidophis* (from Gove, 1979).

Fig. 7.3 Chemoinvestigation among snakes and lizards by tongue-flicks.

Some part of this delivery may involve air-trailing alone and may enable Garter snakes to use TF for prey detection. Blocking a single VN duct resulted in a significant number selecting an air flow with prey odour content presented on the side of the functioning duct (Waters, 1993).

Transfer of scented material dorsally up the duct and into the lumenal cavity is still problematic, if it occurs without recourse to a ciliary current, such as might be exerted by the epithelia of the mushroom body. Suction effects, akin to the mammalian Flehmen mechanism, have been suggested, but not substantiated (Young, 1993). Sub-lingual lamellated processes are raised after tongue retraction, opposing them to the duct apertures and possibly providing a temporary anti-drain wedge. Taste detection at this point is also provided by the receptors on the ridges of the lamellae. Other sites for gustation, apart from the tongue itself, are the general oral surface and the palatal folds flanking the VN entrance area (Schwenk, 1985). Taste as a potential contributor to social chemoreception needs to be evaluated.

Tongue-flicking is conspicuous in many snakes and lizards, whether they have deep or shallow forks to their tongues (Schwenk, 1994). In species with tongue protrusion, with or without oscillations, chemicals are collected by scent contact for delivery to VN duct openings [Fig. 7.3(a)]. The retraction of the tips after sampling, presents their lower (contact) surfaces [Fig. 7.3(b)] to the duct entrances. The tip surfaces seem adapted for transport, since the epithelium covering the tips is the only smooth area; it lacks the papillae and micro features found elsewhere on the lingual surface (Delheusy *et al.*, 1994). TF chemoinvestigation is an almost constant monitoring pattern, which is labile in its basal and its aroused rates, the latter varying over a range of some 30-fold/min (Gravelle, 1980). Although linked with exploration, flick rate has rarely been correlated with other arousal parameters, such as respiratory or heart rate levels. Physiological correlates should assist with the separation of TF in air, from TF with substrate contact, and thus help to define its degree of association with the activation of either system. Several novel environments all raised exploratory TF in a lizard to about the same level, although the TF rate was non-discriminatory (Rybiski, 1995). Nasal (respiratory) intake is assumed to be the principal activator for the MOS providing distance-sensing (Cowles and Phelan, 1958). Responses to conspecific semiochemicals in exudates of male and female fence lizards (*Sceloporus occidentalis*), occurred more readily than to neutral odourants; water did not induce TF arousal (Duvall, 1981). The latency to the onset of air-flicks was taken as a criterion for a model of the MOS-to-AOS sequence for activation of chemoinvestigation, as indeed is often assumed for mammalian arousal. Extracts of female cloacal secretions raised tongue-flick rates in breeding and non-breeding male skinks (*Eumeces laticeps*), after treatment of females with estrogen (Cooper, 1995). Vomerolfactory investigation in those males with presumed high androgen levels (orange heads), exceeded that of low hormone males (tan heads), to cloacal signal fractions; both male types did not change their TF on exposure to non-treated control females.

Other conditions with overt vomerolfaction are in individual discriminations of, for example, own versus strange odour, and kin versus non-kin odour (Cooper *et al.*, 1999). These abilities are widespread amongst lizards, particularly where both sexes defend a territory, or as in mammals, where group-scent enables intruders to be distinguished.

7.1.4 Mammals

"Treat a monkey seriously ... and he'll only flear at you".

(O.E.D., Hickersgill, 1683)

Flehmen (Flehmane or Flehman are variants) is variously defined as a facial grin or grimace, and is a distinctive feature of chemoinvestigation behaviour in the terrestrial mammals. A systematic survey by Schneider (1935) noted its occurrence in many higher mammals; additions to the list [c.f. Figs. 7.6(a) to (h)] now include shrews, wombats, bats and lemurs (Mann, 1961; Bailey, 1978; Gaughwin, 1979; Baxter, 1981). Functional linkage of Flehmen (F) with the presence of a VNO and of a distinct AOB was first made by Knappe (1964), initiating a debate over the details of its chemoreceptive role (c.f. Dagg and Taub, 1970). The perceptive identification of F. as a key element in estrous detection (Fig. 7.7), was made by Estes (1972). Intake of non-volatiles in fluids is considered to be the function of F., it being the primary VN-specific

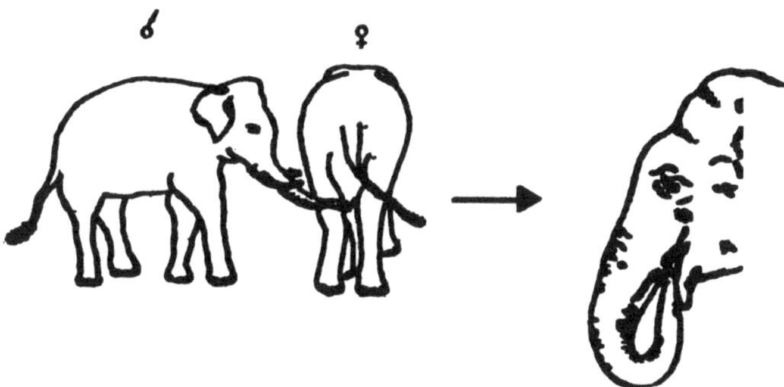

Fig. 7.4(a) Chemoinvestigation with F. in male Indian Elephant: L, Sampling of urino-genital secretions (*in situ*) with trunk-tip (pre-Flehmen) ⇒ R. trunk (and upper-lip) curl, sample placed on palate ⇒ (b. below); (after Eisenberg *et al.*, 1971; after Rasmussen *et al.*, 1982).

sampling mechanism for oral and/or nasal uptake into the lumen (c.f. Figs. 7.5 and 5.7). Signal transfer in mammals employs many and varied means of dispersal (Gorman, 1990). Scent sources and deposition behaviours are ubiquitous, although their social significance is not always apparent (Herrera and Macdonald, 1994). A female Fork-marked lemur carries male scent, deposited [c.f. Fig. 7.2(b)], for intra-specific purposes, in an ungroomable position. Thereafter, the now mobile signal could convey at least male presence, if not "female-ownership" by the territory-holder.

The use of urine as a signal vehicle is prevalent in many mammals, but its usage is not universally associated with chemocommunication. The enigmatic "urine-washing" (Chap. 7 Heading Fig.) whilst common in the New-World monkeys and in prosimians may even be non-communicatory, e.g. grip enhancement. It has not been established as similar in function to urinalysis (cycle monitoring) in non-primates. The transference of voided urine from hind to fore feet has many alternative and plausible attributes. In arboreal species, usages such as locomotor and trail deposition are as likely as is boundary definition (Charles-Dominique, 1977). Squirrel monkeys (*Saimiri oerstedi*) in feral troops show several olfaction-related behaviours; although all occur at low frequency, only close-range genital inspection of females by males was unequivocally reproduction-related (Boinski, 1992). Urinary signalling is established as a major chemocommunication route; as important in mice as it is in elephants. Urine may carry semiochemicals as voided from the bladder, or in combination with reproductive tract/skin-gland secretions. Faecal products are similarly endowed, as pointed out by Darwin (1839), and typical of South American ungulates as well as of Hyenas (Altman, 1969).

Small-bodied groups such as rodents are assumed to sample fluids by vasomotor intake alone [Fig. 5.8(b)] and although rodents seem to lack F., the smallest insectivores and marsupials [Figs. 7.6(a) and (b)] exhibit a version with minimal lip-curl. It may be that F. in many species is similarly inconspicuous and remains undescribed. At the other end of the biomass scale, the largest land mammal displays courtship behaviour with an evident but modified Flehmen. Chemoinvestigation in elephants (*E. maximus*) was described as including "urine-teasing", without equating this to Flehmen (Eisenberg *et al.*, 1971). On close analysis, the tactilely enhanced tip of the trunk (Pl. 7.1) — the finger region with its pore-like openings — dips directly into deposited female urine, or touches urino-genital secretions *in situ*, then it curls ventrally and brings a sample to the palatal duct apertures [Fig. 7.4(b)] leading to the VNO (Rasmussen *et al.*, 1982). The eliciting semiochemical — (Z), 7-dAc (**{9}**., Sec. 3.3) — conveys estrous state, and is itself capable of maintaining F.-responsiveness at similar levels

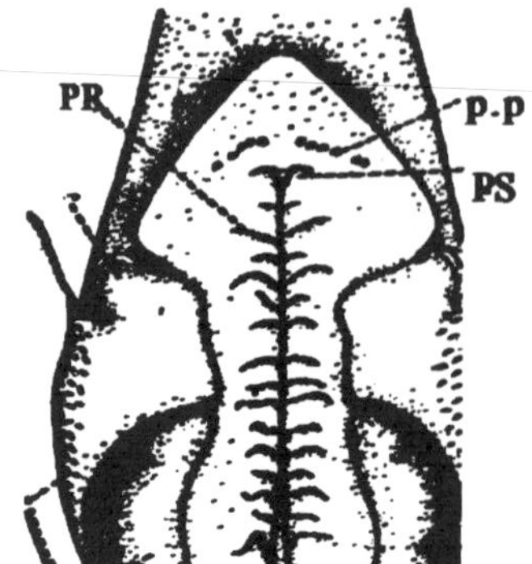

Fig. 7.4(b) Palate of foetal African Elephant: reception area for "finger" of trunk (Pl. 7.1) and for uptake of protein-bound signal complexes. PS, palatine sulcus and pp, palatine pits; PR, ridges (from Eales, 1926).

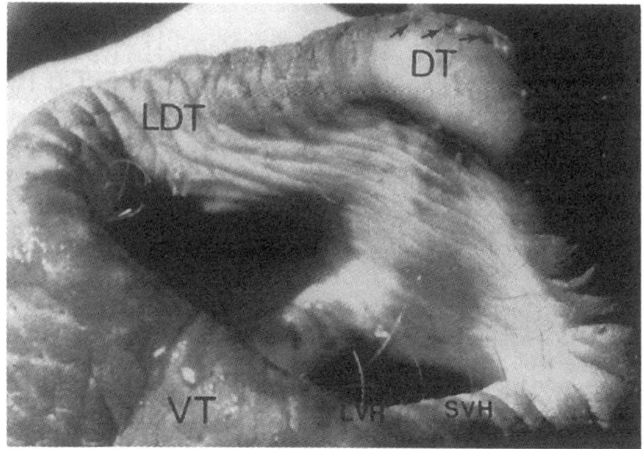

Pl. 7.1 Nasal chemosampling, trunk-tip of Asian Elephant: anterior "finger" region, showing tactile area + pits (DT, arrows). Vibrissal hairs (LVH and SVH), muscular walls (LDT and VT) (from Rasmussen and Munger, 1996).

Fig. 7.5 Chemoinvestigation by sampling scent-mark: F. in male Goat (**a**) nasal contact with o+ urine; and (**b**) neck-stretch and gape (from Ladewig *et al.*, 1980).

to intact urine. The trunk movement is clearly a highly adapted version of the characteristic lip-curl feature of F., and with a modified rhinarium.

Most mammals show one or more of the following elements in F.: jaw opening (gape or yawn) plus an elevated head posture and a temporary stillness; there may also be specific respiratory and tongue movements plus a degree of nostril closure. The repertoire of F. elements occurs in various combinations, and is conspicuous in Ungulate and Carnivore repertoires (Estes, 1972). One or more of the behaviour elements in F., as in Fig. 7.6(a) to (f) may be present in any one species; their separate functions (if any) have not been examined in detail.

The tasting nature of the sequences is often noted in ethograms of e.g. Kangaroos and Antelopes; the insertion of the tongue as well as the rhinarium into the female's urine stream is usually followed by genital nuzzling (Cousin, 1981; Hart and Hart, 1987; Poran *et al.*, 1993). The nature of this pattern suggests that direct contact with the scent source by the rhinarium (Chap. 2), lips and/or tongue is a prerequisite for full F. performance. Alternatively or in addition, prior intake of volatiles may contribute, via MOS activation, to close-contact uptake of non-volatiles (Evans, 1984 and 1986; Wysocki *et al.*, 1985). A model for the combination of F. with the vasomotor pump predicts that F-assisted scent intake is able to present stimulus fluids to the VNd entrance by suction and/or oro-nasal pressure change (Bailey, 1978). A possible device might be that rapid uptake of fluid through the N-Pd creates a modified Venturi-effect in its passage across the VNd entrance towards the nose. Some emptying of the lumen by this means (Pl. 5.2), or by tongue pressure on the palate [Fig. 7.6(e)] may allow the vasomotor apparatus to assist with some or all of the stimulus acquisition sequence [Fig. 5.7(a)]. Thereafter, the AOS exerts autonomic control over alterations to lumenal volume, by a bellows-like action. Undoubtedly the mechanism will vary in detail with morphological constraints. For instance, using their highly modified upper lip, elephants appear to exert pressure on the trunk contents, seemingly to push mucus-bearing odours into the VN duct, a narrow obliquely-angled channel (Rasmussen and Hultgren, 1990, Fig. 1; and Rasmussen, pers. comm., 2000). Uptake in nasal entry species such as the horse, requires nasal contact and direction of the (urine) sample from the upper to lower cavity [Figs. 2.5(d) and 7.6(d)]; the internal configuration of the rostral area presumably facilitates presentation to the VNd aperture. F. is elicited as part of the stallion's monitoring of mares' urine [c.f. Frontisp. and Fig. 6(d)]; its frequency is linked to the volatiles of the estrous cycle stages (Lindsay, 1984; Stahlbaum and Houpt, 1989; Ma and Klemm, 1997). The prominent tongue extension in male tigers [Fig. 6(f)] is problematic.

Confirmation of such adaptations of F. to variations in oro-nasal morphology awaits further recording during the operation of F., for example by refinements of the direct measurements of lumenal changes (Bland and Cottrell, 1989).

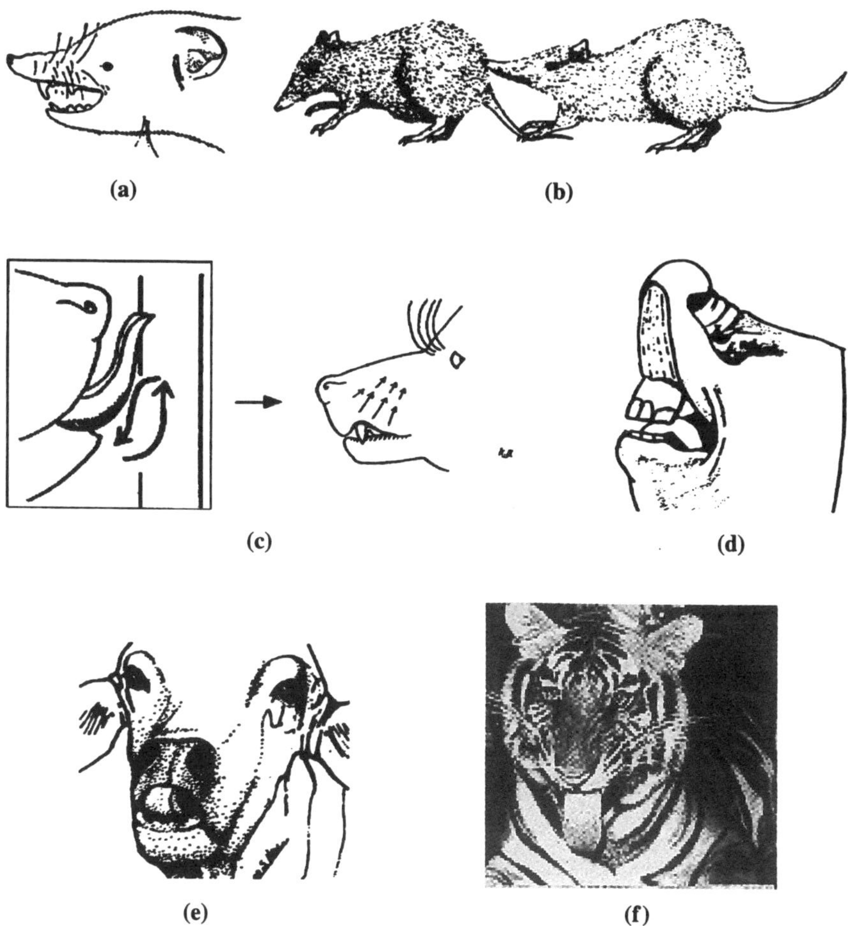

Fig. 7.6 Flehmen (F.) component patterns. **(a)** Gape phase only, in male Shrew *Crocidura hirta* (from Baxter, 1981). **(b)** Genital chemoinvestigation (and "trapping") of female by male Bandicoot (*Perameles nastua*), + lip-curl? (from Stodart, 1966). **(c)** Lapping (licking with extended tongue) by male (or female) Ring-tailed Lemur: intake of labial scent, leading to F. with strong Lip-curl + Gape (from Bailey, 1978). **(d)** Head tilt + Gape + Lip-curl, F. in stallion to estrous urine odour (from Lindsay and Burton, 1983 — c.f. Frontispiece). **(e)** Stimulus-exchange in cattle: the forward tongue-compression stroke (pre-F.?), "rinsing-out" of lumenal fluid [c.f. Pl. 5.2] (from Jacobs *et al.*, 1980). **(f)** Tongue-protrusion in captive male Tiger (*Felis tigris*) (from Brahmachary *et al.*, 2000).

During pre-mating encounters, boar F. is associated with the production of salivary non-volatiles, inducing the receptive posture in the sow (Martys, 1977; Dorries *et al.*, 1995). However, the pig's nose is predominantly a feeding device. Adaptations for rooting include the flattened snout-disc and the cartilages of the external nose which act as a "special supportive system" for its intrinsic muscles (Soucek *et al.*, 1999). Vomeronasal operation in pigs (c.f. Fig. 5.6) requires modified cartilages; those in the snout are no longer independent, but linked to the ventro-lateral nasal cartilage, and may function to transmit passive movement to the VN complex.

From an examination of ungulate usage, F. was proposed as serving principally as a chemosensory pattern for obtaining estrous signals by immediate urinalysis (Estes, 1972). This influential insight directed attention towards F. as an important part of reproductive behaviours. Estes's interpretation appears particularly relevant to monestrous species and to those with restricted but polyestrous breeding periods. In social groups, with or without harem structures, rapid assessment of females' status by the dominant male has obvious advantages (Black-Cleworth and Verberne, 1975; Hart, 1987). The appearance of F. in a "lower" primate (*Lemur catta*) was unexpected [Fig. 7.6(c)] despite its high probability of occurrence, in view of the well-developed VNO and duct system (Evans, 1968 and 1972). The male Ring-tailed lemur continuously assesses the scent marks from individual females throughout the cycle; an extended lapping sequence [Fig. 7.6(c), left] occurs when investigating labial secretions (Evans and Goy, 1968; Bailey, 1978; Dugmore *et al.*, 1984).

Similar constant checking across the cycle occurs in the Spider monkey (*Ateles*), where males directly sniff and nasally contact the females' complex labial folds. The external genitalia lack visual peri-ovulatory alterations, hence usage of the AOS may occur. It is however unestablished in this, as in many other species without overt vulval colour change (Klein, 1971; Hunter *et al.*, 1984).

An adaptive association of Flehmen with fluid-sampling is shown where related modifications to the stimulus-access system, such as anatomical and behavioural anomalies, appear. The F. sequence is inconspicuous and ducts reduced in two African Alcephaline antelopes: the Topi (*Damaliscus iunatus*) and Coke's Hartebeest (*Alcephalus*

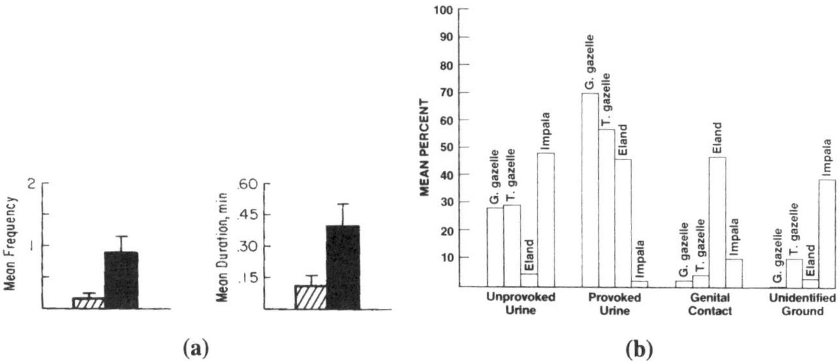

Fig. 7.7 Discrimination by F. (**a**) of estrous, ■ vs. non-estrous, ▨ urine; frequency in feral goats (from O'Brien, 1982); and (**b**) within social groups, species-differences in responsiveness of male antelopes to urinary and/or genital signals (from Hart and Hart, 1987).

buscephalus), despite the presence of a mature VNO (Hart *et al.*, 1988). Both species lack a N-Pp and a N-Pd; in the related Wildebeest (*Connochaetes taurinus*) only nostril licking appears, apparently the only overt remnant of F. components.

A possible explanation could be that tongue use alone sufficiently compensates for suppression of F. sequences to allow uptake and subsequent transfer of urinary signals. It may be that reduced chemo-investigative responses are not readily explained in terms of selection pressures on estrus detection. Possibly these two antelopes are special-needs species which suppress the visual component of F., while maintaining urinalysis in their repertoire. The use of F. as part of some dominance-related, but partly visual, signalling is suggested by its usage in groups of species as distinct as Oryx and rabbit (Pfeiffer, 1985; Black-Cleworth and Verberne, 1975).

A current working hypothesis is that F. and chemoinvestigative behaviour are closely linked, but that it is not restricted to urinalysis by the male. Other signalling functions are indicated by the behaviour of captive and feral groups. Flehmen's role in female ↔ female interactions is clearly of equal value since it is highly correlated with rank in the Sable antelope (*Hippotragus niger*). Dominant females not only perform the highest Flehmen rates, but also show the greatest degree of reproductive synchrony since their F. level peaks at about the estimated time of conception (Thompson, 1995). As seen within male hierarchies, subordinate females in this species rarely perform Flehmen in response to urination by higher-ranking females (Thompson, 1991). Suppression of a behaviour contributing to the (advantageous) synchronicity of parturition could expose low-rank females to fawn loss and/or poorer resource access. Similar social inhibition mechanisms can inhibit ovulation and hence limit the fertility of all but the alpha-female — as in the dwarf mongoose and in the marmoset/tamarin group (Rood, 1980; Barrett *et al.*, 1993). Indeed a limited role for the AOS in mediation of the suppressive influence of the breeding female (Chap. 5) has received experimental support (Abbott *et al.*, 1988). The presence of the dominant female within family groups of Callitrichid monkeys exerts a multisensory influence on subordinate females. MOS and/or AOS involvement is proposed as the primary route for the detection of ovulation (Abbott *et al.*, 1990; Ziegler *et al.*, 1993; Carlson *et al.*, 1999; and Chap. 5). As yet, no description of typical mammalian F. patterns has appeared for any South American monkey, although close range naso-labial inspection, plus rapid tongue-flicking, does occur and may well contribute to cycle-phase analysis (Epple, 1986; pers. comm., 2000; Smith *et al.*, 1998).

The determinants of F. have been subject to experimentation mostly in field and captive studies of ungulates. These support the expected association between the frequency and occurrence of Flehmen and the seasonality of reproduction. The elicitation of F. can also depend upon the social context; presentation of urine or other stimuli alone may not produce consistent displays. When conspecific urine was tested out of context (i.e. no female present) in male Black-tailed deer, there was no discrimination between urine from individual adult males or between urine from estrous/non-estrous females (Altieri, 1980). Correlation of male endocrine status in reindeer (*Rangifer tarandus*) showed that the elevation of testosterone during rut and the duration of F. elicited by female urine was coincident; F. bouts during rut were twice as long following exposure to adult female urine as to that of immature females (Mossing and Damber, 1981).

During development, pre-pubertal animals without sexual experience are unlikely to use F., since it presumably requires both hormonal and experiential accumulation. Longitudinal studies are scarce, but Sable antelopes show a gradual rise in the appearance of the F. elicitation to adult urine up to two years of age (Fig. 7.8). Surprisingly, young fawns (male and female) at about two-months-old consistently gave F., thereafter a significant sex difference in the F. frequency appeared (Thompson, 1995).

The influence of gonadal hormones on prepubertal animals suggests some steroidal sensitivity in adults with regard to F. elicitation. Young male sheep are induced to perform F. in response to exogenous T and to 17-β-estradiol; F. in female red deer is also sensitive to T injections (Parrott, 1978; Fletcher, 1978). Sex differences can interact with the hormonal state where social conditions vary. Female cats (intact) display F. to urine marks only in the absence of males; testosterone propionate induced F. in spayed females towards estrous females (Verberne, 1976; Hart and Leedy, 1987), whereas an ovarian hormone (estradiol) failed to elicit F. to males (intact, and sexually inactive), presumably indicative of social inhibition overriding steroid facilitation.

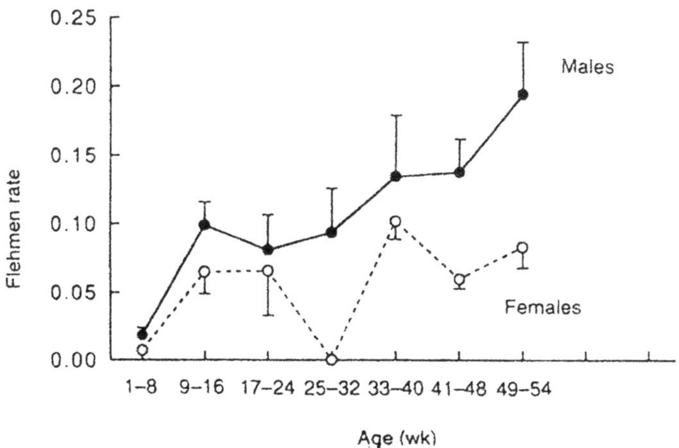

Fig. 7.8 Sex differences in F. frequency (/hr) with age: birth to one year, in captive Sable Antelope *Hippotragus niger* (from Thompson, 1995).

Centrally mediated influence over F. expression is indicated for castrate horses (geldings); those receiving GnRH + T gave more frequent and longer F. to estrous mares than subjects given T only (McDonnell *et al.*, 1989). The effects of hypothalamic lesions, which abolish copulatory patterns, also support these findings; F. in cats is untouched, suggesting that its continued expression is mediated by non-mating neural pathways (Hart and Leedy, 1987).

A correlation between hormone-dependent behaviour (aggression) and F. frequencies was found for captive cow/calf only herds of the Scimitar-horned Oryx, but only during one month of a six-month study. However, in all-female groups of this oryx, there was no rank correlation, indeed dominants both gave and received significantly less F. than those of lower rank — mostly the juveniles (Pfeiffer, 1985). The inter-female monitoring of cycle status with F. suggested that, as with Sable antelopes, conception/birth synchrony might be present and contribute to calf survival during the formation of crèches. The usage of F., even in antelopes with strict harem systems, is again not necessarily restricted to males. Female captive Blackbucks, which show such systems, exceeded males in urine sampling, with no differential display of F. by and to males and females (Schmied, 1973).

Flehmen in all species so far studied is stereotyped in execution, although as noted, not all typical patterns are part of each species' repertoire. Its function(s) may well vary with species requirements, as with the modifications in certain antelopes discussed above (Hart *et al.*, 1988). F. is not always dimorphic and can be influenced by social context: the licking/lapping sequence aids discriminations among female as well as male lemurs [Fig. 7.9(a)]. The contribution of F. to fertility is bound up with the role of the AOS in reproductive patterns. Not all species with a full complement of accessory olfactory structures have been scrutinised for F., nevertheless it is rarely absent, even if modified, in species with VNO/AOB presence. Its presence has not been reported in species without a functional AOS, as required by Knappe's hypothesis (Knappe, 1964).

Those species in which a urinary component is necessary, predominant or highly likely as a contributor, are clearly of most relevance to AOS function. The importance of the linkage between Flehmen and the AOS is the most likely of the functional associations. It has been stressed by numerous commentators that the roles and relative importance of the olfactory systems cannot be unequivocally established from unsupported behavioural observations, however informative, of feral or captive animals, although elephants may prove the exceptional case. Even behaviours incorporating licking of a scent source, while suggestive, do not permit inferences on activation of the organ unless uptake has already been shown to follow tongue and or nasal contact in that species. Into this category would fall observations of licking, F. and the occurrence of VN uptake (Chap. 5) by goats, guinea pigs and mouse-lemurs. These species all employed urinalysis in their assessment of estrus or other status; scent-gland semiochemicals are as effective in elicitation of F. In seasonally polyestrous species, rapid discrimination of female receptivity is important (Evans and Goy, 1968; Pereira, 1991). In strict monestrous species it is a necessity, as with Colombian Ground squirrels, in which mating occurs immediately on emergence from winter dens (Harris and Murie, 1984). Despite staged observations, during males' chemoinvestigation of neutral objects, only the occurrence of sniffing was evident and discriminatory (of proestrus); while licking to the vulval smears presented was present but infrequently seen clearly at 25 m distance. Unless a pattern strongly associated with Flehmen [Figs. 7.6(a) to (f)] can be unequivocally identified, then AOS involvement cannot be given credence.

7.2 INTER-SPECIFIC INTERACTIONS

Within cryptic and nocturnal species, and particularly in those with overlapping ranging (and often resource) needs within an area (sympatric species), olfaction is considered an important genetically isolating factor. The relative contribution of AOS versus MOS signals are again not simple to disentangle, since few relevant studies consider heterospecific signalling, and an even smaller number to sufficient analytic depth.

7.2.1 Predator/Prey

Amongst amphibia, chemosignalling is most prominent as part of mate location, courtship and mating (Chaps. 2 and 3). Social usage of scent marks within and between species in the promotion of territorial functions seems variable. The most probable correlations are with habitat type (aquatic, semi-aquatic or terrestrial) and reproductive seasonality (Tristram, 1977; Dawley, 1992 and 1995). Some weak discriminations by male salamanders of substrate marks or faecal deposits suggest avoidance of other males in some, but not all contexts (Jaeger and Gergits, 1979). Males avoid strange male conspecifics only in the absence of female signals; strange female (*Plethodon cinereus*) congeners showed some labile avoidance behaviours interpreted as "lack of preference for own scent". The simplicity of social structures can be seen as the basis for the infrequent amphibian usage of isolating chemosignals. A role in predator detection and avoidance at present cannot be excluded.

As their primary chemoinvestigatory mode, reptiles use vomerolfaction; even during underwater hunting, TF is maintained. There are, however, examples of visual ↔ olfactory synergy in the elicitation of TF. In garter snakes, stimuli from prey which display warning (aposematic) colouration provides such a case (Terrick *et al.*, 1995). TF rates rise significantly and prey selection delay drops in choice situations of "warning versus neutral" prey. Such avoidance behaviour is present in all vertebrates but not necessarily with a comparable linkage. Scleroglossan lizards show in their chemoinvestigative behaviour TF frequencies which are highly correlated with the morphological complexity of the VNC and the size of AOB; adaptive complexes suited to active foragers and trail-followers alike. In contrast, ambush-foragers like crotaline vipers and some boids (pythons) seem to show a reduction in VN participation, with infrequent or low tongue-flick rates (Gabe, 1976 and 1990). Where hunting is dominated by air-borne cues as in trailing species, there is an expected shift to dependence upon the MOS, as in lizards (c.f. Simon, 1983). Garter snakes could not locate earthworm odour by the AOS alone (Chap. 5) when given a choice of air-stream odours (Halpern *et al.*, 1997). However, when allowed contact with a trail scent, prey cues were followed correctly, but not after VNd blockage (Waters, 1993). Active hunting then, involves both systems for optimal success, whereas predator avoidance can be solely reliant on the AOS. In eight species of pit-vipers, prey to an ophiophagous (snake-eating) predator, avoidance depends upon recognition of the skin secretions "integumentary signals" of the common Kingsnake (*Lampropeltis getula*) (Miller and Gutzke, 1999). More remarkably, its presence is detected primarily by vomerolfaction, as inferred by the behaviour of pit-vipers with sutured VN ducts, when placed with an accessible Kingsnake.

Location of conspecifics in the presence of sympatrics, whether closely related or not, is another skin-lipid based response. Western ribbon snakes (*Thamnophis proximus*) could not locate own-species scent without VNO access; sympatric scents were not recognised (Graves, 1991).

7.2.2 Sympatric

Mammalian speciation presents chemoreception studies with considerable problems; rarely can semiochemicals be shown to act as the major element in the process. The presence of some shared chemosensory characteristics (dimorphic signals) in sympatric North American shrews is illustrative. The major signal source is the flank gland which secretes in adult males during seasonal reproductive activity (Hawes, 1976). The odours emitted are reported to be distinguishable by the human nose — "musky" in *Sorex vagrans* and "acrid" in *S. obscurus*. Whether the semiochemicals produced are contributory to mutual avoidance and niche adaptation is questionable without, e.g. preference testing. Intra-specific usage in this genus is more clearly indicated in *S. araneus*, where additionally the non-estrous females show considerable flank gland activity.

Heterospecific chemical signalling appears in studies of the temporal segregation of closely related species. Odour exposure was most effective in the taxon with labile features to its activity rhythm. Friedman *et al.* (1997) found such interactions in Golden Spiny mice (*Acomys russatus*). Exposure to odours from *A. cahirinus*, a co-existing sympatric, forced *A. russatus* to be active at less preferred times. Food niche specialisation, as well as differential activity rhythms, apply to co-existing species with narrow (plant) preferences (Petter and Peyrieras, 1970). All three species of Bamboo lemur (*Hapalemur*) overlap in selection of their eponymous food-plant (85% commonality) but show micro-selection of preferred bamboo parts (Wright and Randrimanantena, 1989). The probability of scent-based discrimination existing among this group of species is supported by the comparability of their chemosignal production. All three use a dual gland system in males to construct the composition of their scent marks, which are indicated as partially an inter-specific signal source within the genus. Captive lemurs without visual contact, but in proximity, selectively over-marked hetero-specific marks *H. aureus* versus *H. griseus* and *H. griseus* versus *H. simus* (Evans, unpubl.). The Gray-bamboo lemur is also sympatric with a fourth dual-gland species, the Ring-tailed lemur. Heterogeneric preference is again shown by these two lemurs; “strange”-species scents are preferentially licked and over-marked (Evans, 1986).

As mentioned earlier (Chap. 3), the chemosensory role in speciation of blind Mole-rats (*Spalax* spp.) is considerable. Indeed, it extends to solely chromosomally-distinguished (2n = 58 and 2n = 60) sibling species (Todrank and Heth, 1996). Of a subterranean superspecies group (*S. ehrenbergi*), four sibling types show odour discriminations amongst each other which reflect their genetic relatedness (Heth and Todrank, 2000). Within-sex testing revealed that blind Mole-rats preferred the species-odour of urine from species most similar to themselves. The process of genetic drift appears to be accelerated by the ability to perceive urinary distinctions between other species of varying relatedness. Within the superspecies group, individual discriminations appeared to derive from variance with respect to the degree of deviation from an ancestral species-type. A matching or comparator mechanism of self with non-self patterns of signal components could be at work allowing an estimation of the extent of divergence to be tolerated.

The ability to compare semiochemical patterns and to assess degrees of similarity is a likely mechanism underlying the establishment, and eventually the maintenance of genetic divergence. Amongst Tamarins (*Saguinus* spp.) and Marmosets, both for within- and between- (related) species distinctions, the complexity of their scent marks provides a discriminatory mechanism (Epple *et al.*, 1987; Smith *et al.*, 1997). Mixed species groups of these monkeys are enabled to co-exist, to minimise resource competition, and more importantly, to maintain the genetic distances produced by prior geographic separation.

7.3 SOCIO-SEXUAL INTERACTIONS

7.3.1 Modal Interactions

Ultra-sound emissions typically occur when male rodents are exposed to female odours or altricial neonates to maternal sources (Whitney, 1974; Conely and Bell, 1978). Without the VNO, sexually inexperienced male mice do not utter emissions at ultra-high frequencies (UHF), whereas those with prior experience vocalise after VN-x, as discussed above (Chap. 5). Female mouse urine contains a unique UHF-eliciting component which is non-volatile but ephemeral (Sipos *et al.*, 1995). The signal is degraded by oxidation and disappears within 15 to 18 hours of deposition. Direct contact with freshly voided urine must occur before males will vocalise (sexually experienced or inexperienced). At least one of the olfactory systems is needed for UHF to be elicited by fresh urine; complete deafferentation abolishes the response (Sipos *et al.*, 1993). Exposure to females permits UHF to be elicited by other than chemical cues (Labov and Wysocki, 1989). Nocturnal or cryptic species conceivably use ultrasound to advertise male presence; whether this is to deter other males or assist with female location is unclear.

UHF also plays a prominent role in the ability of rat neonates to elicit maternal licking of their perineal skin; without such tactile input, defecation fatally fails. The chemosignal appears to be a single (mono-component) semiochemical as well as being a VNO-mediated stimulus (Brouette-Lahlou *et al.*, 1992 and 1999). The dodecyl propionate in preputial gland secretion was found to orientate the maternal anogenital licking of the pup (Chap. 4), and was initiated in response to its UHF distress calls (Brouette-Lahlou, 1992).

The belt-and-braces sequence of sound + chemical ⇒ tactile response is explained by the necessity of perineal stimulation. In its absence defecation cannot begin, with lethal consequences. By 12 days, the pup's UHF response discriminates male and female adults; calls are given only after removal of anaesthetised females, but not on male removal; anosmia abolishes the discrimination unless an active female is present

— partial effects were not tested (Shair *et al.*, 1999). Early learning of non-olfactory cues clearly aids even pre-weaning pups.

7.3.2 Individual

The majority of vomerolfactory effects discussed relate to intra-specific patterns, most concerned with social discriminations. Individual and or group membership, hierarchical status — often aggression-related, are among many non-sexual but socially indispensable elements. Social systems with evident female dominance are not infrequent among

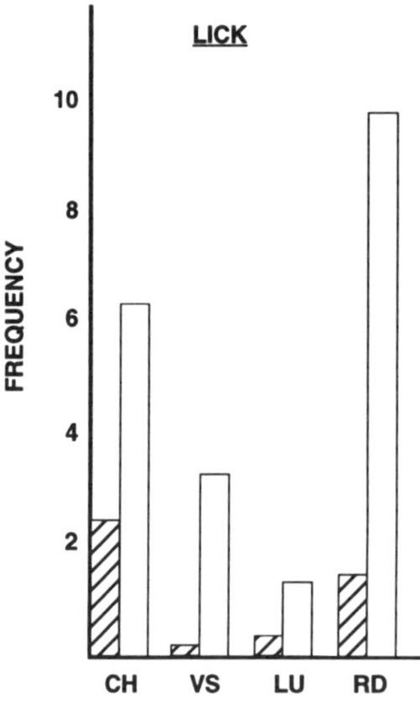

Fig. 7.9(a) Discrimination of labial secretions by female Ring-tailed lemurs: differential VN-related responses (licks, c.f.; 6,c/5 min) to own, ▨ vs. strange-female, □ scent (from Dugmore, 1984).

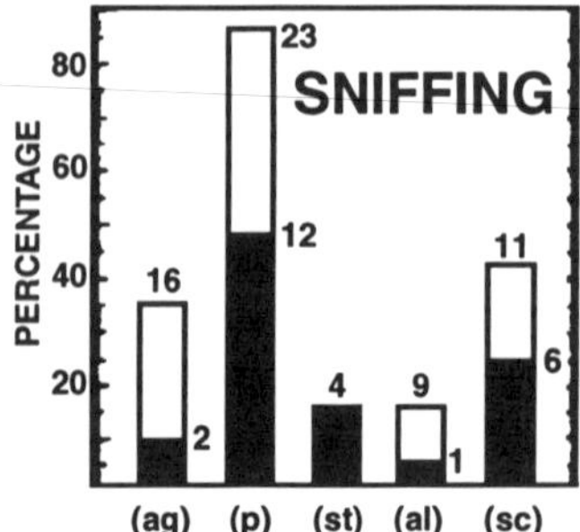

Fig. 7.9(b) Chemoinvestigation among Viverrids (c.f. Chap. 3, Heading Fig.): localised, site-specific investigation of skin-gland complexes in *Genetta*; ano-genital (ag), perineal (p), scrotal (st), anal (al), and sub-caudal (sc). [*nos.* = *sniffs* (sec.), ■ = % of all observations, frequency and duration.] (from Wemmer, 1977).

prosimians, as is functionality of the AOS (Pl. 2.1B) (Izard, 1990; Evans and Schilling, 1995). Labial secretions deposited by females are frequently and assiduously licked by other females [Fig. 7.9(a)]. Other than during the very brief (~12 hours) estrus, vaginal secretions do not contribute to the female signal (Evans, unpubl.). The mixed (apocrine/ sebaceous) exocrine products which accumulate on the labial folds allow individual discriminations, without the necessity of directly contacting the sender (Mertl, 1975; Dugmore, 1984; Dugmore and Evans, 1990).

In a multisignalling context, there can be an extensive array of scent sources elaborated on even one localised part of the body surface [Fig. 7.9(b)]. The distribution of sniffing in a semi-social carnivore (Genet) clearly shows that all the glands, etc. are given varying degrees of attention (Wemmer, 1977). Assignment of even approximate communication values and functional import to each distinct source is not a trivial task, let alone the division of the inputs to each system. Combinatorial emissions may well also occur, i.e. two or more distinct scent sources are emitted separately on some occasions and jointly in other contexts. Apocrine and sebaceous mixing perhaps for "fixation"

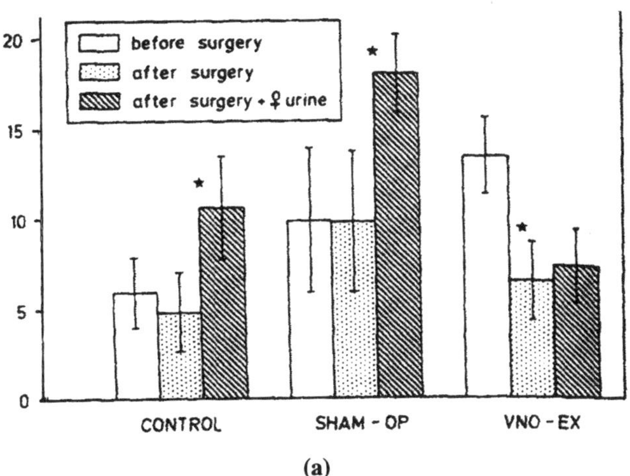

(a)

Fig. 7.10 Vomeronasalectomy: effects on chemocommunication. **(a)** Inhibitory effect of VN-x on urine-induction of scent marking in male Gerbils: frequency/5 min (from Probst, 1990).

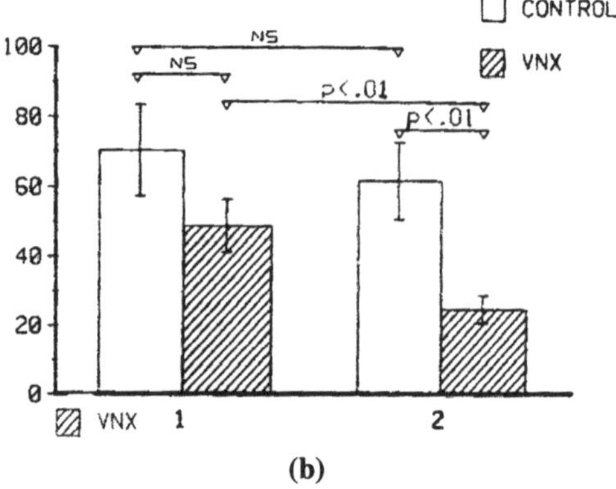

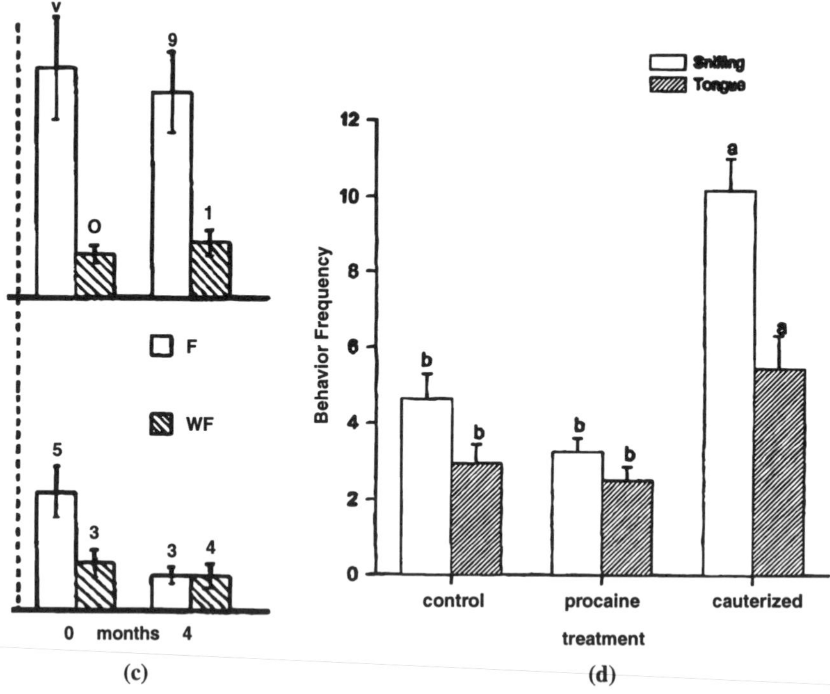

Fig. 7.10 (*Continued*) **(b)** VN-x inhibits response to female urine by male guinea-pig; X-2 sequential trials (duration, sec. x ± s.e.) (from Beauchamp *et al.*, 1982). **(c)** Inter-strain (domestic vs. wild) discrimination by male domestic guinea pig of female urines [sequential testing: sec/4 min, ± s.e., Ss above each bar; intact/sham-VN-x. above]. WF = wild female, F = domestic female (from Beauchamp *et al.*, 1982). **(d)** Effects of VN-x on maternal chemoinvestigation: ewe responses to lambs, including tongue-manipulation of palate [c.f. Fig. 7.6(d)], procaine = MOE inhibition, "a" sign. versus control (from Booth and Katz, 2000).

of volatiles, occurs in shrews, hyenas, and in Ring-tailed lemurs, during inter-male combat or courtship (Dryden and Conaway, 1967; Broman, 1990; Evans and Goy, 1968).

In reproductive contexts, chemosignalling is generally universal across the major groups. Induction of scent dispersal contributes to mating itself and to its consequences in infant survival (Fig. 7.10).

The consistency of the effects noted above and in Chap. 5 which modulate reproduction and involve the operation of the AOS, are unlikely to be wholly artifacts of captivity. Nevertheless, it needs to be shown which of the influences on male and female fertility have relevance to natural populations (Fig. 7.12). As mentioned, very few experiments on free-living social mammals have been reported, since the logistical problems of stimulus manipulation and control are formidable. A semi-feral population is an acceptable substitute, and provides some means of testing assumptions on the relevance of findings on caged laboratory-bred rodents.

7.3.3 Populations

A few studies have directly implicated urinary signals in the neurocrine effects which influence population growth and mediate the effects of crowded (high-density) populations (Chaps. 3 and 5). Prairie voles

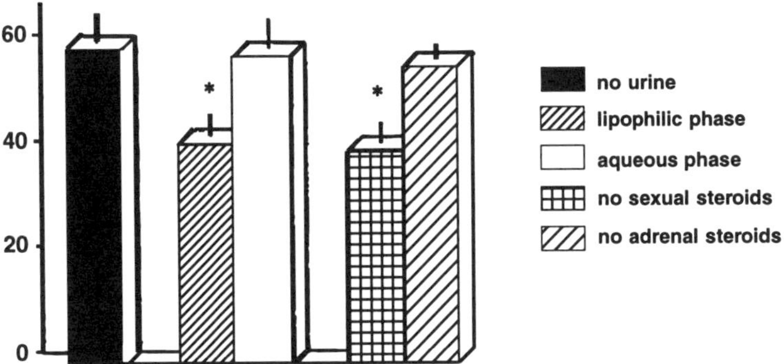

Fig. 7.11 Inhibitory effect of *alpha*-Male urine on plasma Testosterone [T. ng/ml]. Exposure to urinary fractions in isolated (subordinate) male Mouse lemurs (N = 10) vs. castrate/adrenalectomised Ss, *p < 0.001 vs. control (from Perret, 1995).

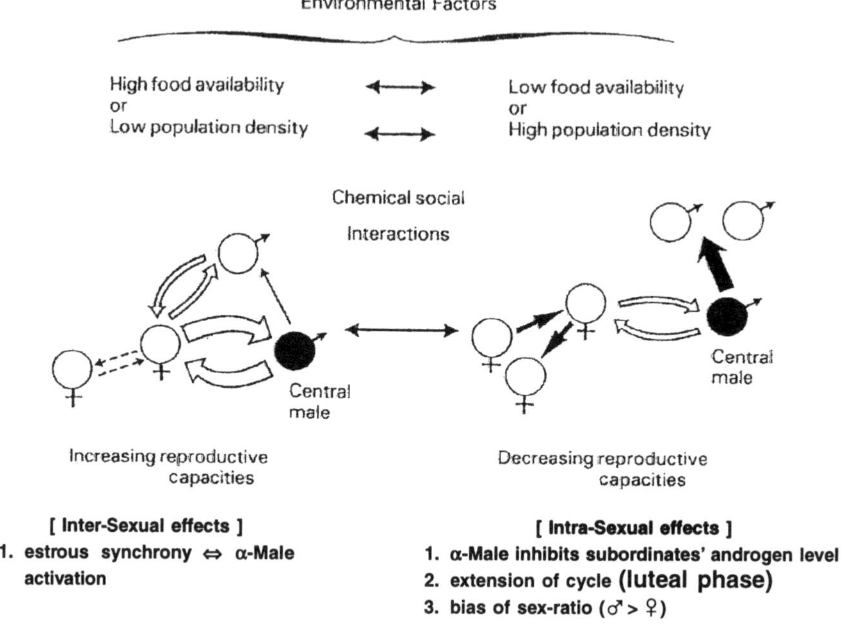

Fig. 7.12 Urinary pheromones and *Microcebus* populations: reproductive influences on Mouse-lemurs. Left: weak intra-sexual effects, Right: weak inter-sexual effects (c.f. Fig. 7.11); for interactions with photoperiod effects, see text (after Perret, 1992 and 1995; Schilling *et al.*, 1984).

(*M. onchrogaster*) were examined for the effects of signals produced within a semi-natural habitat. They were allowed to breed in outdoor pens until their density was similar to natural populations. In this situation, critical exposure to male chemosignals was not enforced by proximity. A strange male was introduced to a male/female pair, or the stud (original) male was replaced by a strange male vole; in the control condition, the mated pair was left undisturbed (Heske and Nelson, 1989). From the records of the birth dates of subsequent litters, pregnancy interruptions were found in both experimental conditions. The female was affected when she could have either avoided or repelled a strange male, and when male ↔ male fighting occurred. A degree of vomeronasal contribution is again assumed, but cannot be quantified.

Similarly, feral house mouse populations living within a restricted area (clover-leaf road junctions, or "highway-islands") were sampled for urine, and captive juvenile females exposed to standardised volumes of urine from these and from control (low-density) groups. Chemosignals from "crowded" adult females prolonged puberty in laboratory-bred conspecifics (Massey and Vandenbergh, 1980; Vandenbergh, 1989). The benefits from this may concern both sender and receiver, since the temporary suppressive effects allow for an adaptive interaction between the timing of puberty and social conditions (Coppola, 1986; Coppola and Vandenbergh, 1987).

Urinary semiochemicals not only modify endocrine reproductive events [Fig. 5.3(a)], but also demonstrate a probable vomerolfactory perspective on space usage. The responses of adult female mice influence their breeding success through choice of nest sites. The preferences expressed to strange/familiar combinations of female odours were consistent with avoidance of novel urine-marked sites, whether from pregnant or non-pregnant females (Hurst and Nevison, 1994). The authors suggest that female urinary cues provide occupancy information and permit spacing of nest sites as their primary function, rather than the expected reproductive (priming) effects. Further analysis of the relevance of the chemical composition of urine marks and of the value of the protein (MUP) content gave insights on the importance of renewal of male marks. Wild-caught male mice were tested with intact strange urine, or with samples whose VNPr complex was separated. The treated urine contained non-bound active ligands (thiazole and brevicomin) whose signal content (Chaps. 3 and 5) denotes status, but are usually available to other males as a MUP-bound complex (Hurst *et al.*, 1998). The results demonstrated avoidance of fresh urine immediately on deposit and up to 1.0-hour-old; by 24 hours the presence of displaced signal compounds no longer induced avoidance. The "fade-out" time of the dominance-related ligands points to a key role for the VNPr in maintaining male presence by extension of the release time of the socially relevant compounds. Without some such device, territory-holding males could not advertise their defended area without an unproductive replacement regime, in order to renew rapidly evaporating odours. Hurst and colleagues also speculate that a further advantage lies in the nature of a mark compared with the physical presence of the territorial dominant. Intruding males may be able to estimate not only age-of-mark but will also equate the volatile signals with probability of attack, hence minimising conflict by non-contact evaluation. The density of marks and their content presumably supply a sufficient concentration of ligands to convey a “dominant male present” message to the advantage of the signaller. Competitive scent marking is a potential mechanism to allow mates to discriminate between individuals of apparently varying quality. Odour content could allow potential mates to avoid individuals of low status (reproductive risk), poor health (parasite-burden) or unsuitable genotype (t-allele presence) (Coopersmith and Lenington, 1998; Rich and Hurst, 1998).

Amongst primate societies, those with a nocturnal habit seem likely to contain the most elaborate chemocommunication content (Charles-Dominique, 1977; Evans and Goy, 1968). The social communication network in Mouse-lemurs (*Microcebus* spp.) is one in which a strong case has been established for an AOS contribution to population dynamics. It is one of those rare primates in which a plausible model can be applied to the interactions found within natural populations. Urinary signals in the Gray-mouse lemur (*M. murinus*) selectively activate the AOB via a prominent VNO (Schilling *et al.*, 1990). Their effects are unusual in that they are manifested as male-on-male inhibition, as well as the expected female-to-male arousal. These exocrine to endocrine linkages are mediated by an interaction of social status, photoperiod and the pressures of local resource conditions. A summary of the socioecological structures operating under varied environmental conditions is given in Fig. 7.12. As a successful and wide-ranging genus, Mouse-lemurs adjust to extremes of population status (Martin, 1990). The interplay of the effects of urinary chemosignals with seasonal daylength give rise to a duality of outcomes for subordinate males. Androgen levels rise seven times when induced by long-days (>12 hours), but exposure to dominant male urine prevents testicular activity (Perret, 1992). The lipophilic fraction appears to mediate this effect; again an unusual feature is its dependence on adrenocortical integrity. Urine from castrate males still retains inhibitory effectiveness (Fig. 7.11); whereas in males with cortisol suppression, the urine produced was now disinhibitory. However, dominant male urine allowed the arousal of subordinates' testosterone also under short-day conditions (Perret and Schilling, 1995). This appearance of a contrasting chemosignal effect, i.e. removal of suppression, may allow the dispersal of the now "effective" males under favourable conditions (Fig. 7.12, left-schema). Dominance effects are minimal and reciprocal male arousal and estrous synchrony promote breeding success. A similar linkage of seemingly adaptive pheromonal and reproductive effects is seen in the results of female-to-female urinary exposure (Perret, 1996). Increasing duration of exposure to strange female urine swings the bias of sex ratio at birth from 30% : 70% to 70% : 30% (males : females). This pre-conception alteration to fertility appears to vary with inter-sexual competition, as does interference with the luteal phase duration (Fig. 7.12, right-schema). Alterations to the levels of prolactin and other pituitary gonadotrophins are produced by the detection of semiochemicals, the resulting input changing the levels of hypothalamic neurotransmitters (Larriva-Sahd *et al.*, 1993; Fabre-Nys *et al.*, 1997; Halem *et al.*, 1999). The extent of these effects is further mediated by the state of the photoperiod at the time of chemosensory exposure.

The sensitivity of Mouse-lemurs, as with house mice, shows that complex accessory olfactory (urinary)-related responses convincingly demonstrate the ability of a species to adjust reproductive output to social and environmental conditions.

SELECTED BIBLIOGRAPHY (Further Reading)

Albone E. (1984). *Mammalian Semiochemistry.* John Wiley & Sons, Chichester, p. 360.

Beauchamp G.K. and Bartoshuk L., eds (1997). *Tasting and Smelling: Taste and Smell,* 2nd ed. Academic Press, New York, p. 256.

Beidler L.M., ed. (1971). *Handbook of Sensory Physiology: Chemical Senses Pt.1, "Olfaction".* Springer, Berlin, Vol. 4, p. 518.

Birch M.C., ed. (1974). *Pheromones.* Elsevier, Amsterdam, p. 495.

Breipohl W., ed. (1982). *Olfaction and Endocrine Regulation (ECRO Symposium,* **4***),* IRL Press, London, p. 409.

Breipohl W., ed. (1986). *Ontogeny of Olfaction.* Springer, Berlin, p. 268.

Brown R.E. and Macdonald D.W., eds. (1985). *Social Odours in Mammals.* Clarendon Press, Oxford, Vols. 1 & 2, p. 506.

Chadwick D., Marsh J. and Goode J., eds. (1993). *The Molecular Basis of Smell and Taste Transduction.* John Wiley, Chichester, p. 287.

Demski L.S. and Schwanzel-Fukada M., eds. (1987). The Terminal Nerve (*Nervus Terminalis*) — structure function and evolution. *Ann NY Acad Sci* **519**, p. 531.

Doty R.L., ed. (1976). *Mammalian Olfaction, Reproductive Processes and Behavior.* Academic Press, New York, p. 344.

Doty R.L., ed. (2002, in press). *Handbook of Olfaction and Gustation*, 2nd ed. M. Decker, N.Y. & Basel.

Farbmann A.I. (1992). *Cell Biology of Olfaction.* Cambridge University Press, p. 282.

Finger T.E., Silver W.L. and Restrepo D., eds. (2000). *The Neurobiology of Taste and Smell,* 2nd ed. Plenum, New York, p. 479.

Getchell T.V., Doty R.L., Bartoshuk L.M. and Snow J.B., eds. (in press). *Smell and Taste in Health and Disease*, 2nd ed. Raven Press, New York.

Halasz N. (1990). *The Vertebrate Olfactory System.* Akad. Kiado, Budapest, p. 281.

Hara T.J., ed. (1992). *Fish Chemoreception.* Chapman & Hall, London, p. 373.

Johnston J.W. Jr., Moulton D.G. and Turk A., eds. (1970). *Advances in Chemoreception* **1**: *Communication by Chemical Signals*. Appleton-Century-Crofts, New York, p. 412.

Kappers C.U.A., Huber G.C. and Crosby E.C. (1936). *The Comparative Anatomy of the Nervous System of Vertebrates*. Hafner (reprint, 1986), New York, Vol. 3, p. 379.

Laing D.G., ed (1989) *Perception of Complex Smells and Tastes*. Academic Press, San Diego, p. 322.

Margolis F.L. and Getchell T.V., eds. (1988). *Molecular Neurobiology of the Olfactory System*. Plenum, New York, p. 379.

Moulton D.G., Turk A. and Johnston J.W. Jr., eds. (1975). *Methods in Olfactory Research*. Academic Press, New York, p. 497.

Murphy C., ed. (1998). Olfaction and Taste XII. *Ann NY Acad Sci* **855**, p. 872.

Müller-Schwarze D., *et al.* (various; 1977 – on), – to Marchlewska-Koj A., *et al.*, eds. (2001). *Chemical Signals in Vertebrates* **1–9**. Plenum & Others, Oxford, New York, London.

Negus V.E. (1958). *Comparative Anatomy and Physiology of the Nose and Para-Nasal Sinuses*. Oliver & Boyd, Edinburgh, p. 402.

Ohloff G. (1994). *Scent and Fragrances: The Fascination of Odors and their Chemical Perspectives*. Springer, Berlin, p. 238.

Parker G.H. (1922). *Smell, Taste and Allied Senses in the Vertebrates*. Lippincott, Philadelphia, p. 192.

Pfaff D.W., ed. (1985). *Taste, Olfaction, and the Central Nervous System*. Rockefeller University Press, New York, p. 346.

Ritter F., ed. (1979). *Chemical Ecology: Odour Communication in Animals*. Elsevier, Amsterdam, p. 427.

Roper S.D. and Atema J., eds. (1987) Olfaction and Taste IX. *Ann NY Acad Sci* **510**, p. 747.

Schild D., ed. (1990). *Chemosensory Information Processing*. Springer, Berlin (NATO/ASI, *Cell Biol* **39**), p. 403.

Sebeok T.A., ed. (1977). *How Animals Communicate*. Illinois University Press, Urbana, p. 1128.

Serby M. and Chobor K., eds. (1992). *The Science of Olfaction*. Springer, Berlin, p. 590.

Simon S.A. and Nicolelis M.A.L., eds. (2001). *Methods in Chemosensory Research*. CRC Press, Boca Raton, p. 328.

Spielman A. and Brand J., eds. (1995). *Experimental Cell Biology of Taste and Olfaction*. CRC Press Inc, Boca Raton, p. 437.

Steiner J.E. and Ganchrow J.R., eds., (1982). *Determination of Behaviour by Chemical Stimuli* (ECRO Symposium, Vol. 5), IRL Press, London, p. 287.

Stoddart D.M. (1980). *The Ecology of Vertebrate Olfaction.* Methuen, New York, p. 234.

Stoddart D.M., ed. (1980). *Olfaction in Mammals.* Symposium of the Zoological Society of London, Academic Press, London, p. 368.

Stoddart D.M. (1990). *The Scented Ape.* Cambridge University Press, p. 286.

van Toller S. and Dodd G., eds. (1988). *Perfumery: The Psychology and Biology of Fragrance.* Chapman & Hall, London, p. 268.

Vandenbergh J.G., ed. (1983). *Pheromones and Reproduction in Mammals.* Academic Press, London, p. 344.

Wolstenholme G. and Knight J., eds. (1970). *Taste and Smell in Vertebrates.* J & A Churchill/Ciba Foundation, London, p. 402.

Zuckerkandl E. (1887). *Das periphere Geruchsorgan der Saugethiere, eine verglichenden antaomische studie.* Enke, Stuttgart, p. 116.

Other Sources

1. OR Databases:

   http://ycmi.med.yale.edu/senselab/ord/.
   See Skoufos *et al.* (2000) for nomenclature; see also Glusman G. *et al.* (2000). *Mamm Genome* **11**, 1016–1023 and (2001) *Genome Res* **11**, 685–702 for Human Olfactory Receptor Data Exploratorium [http://bioinfo.weizmann.ac.il/HORDE]

2. Homepage Sites for Professional Bodies:

   American Chemoreception Society:
   http://www.achems.org

   European Chemoreception Research Organisation:
   http://www.ecro-online.org

   Japanese Association for Smell and Taste Sciences:
   http://epn.hal.kagoshima-u.ac.jp/JASTE

REFERENCES

Adamopoulos D.A., Kontogeorgos L., Vassilopoulos P., Kapolla N., *et al.* (1992). Effects of induced peripheral anosmia on gonadal maturation in prepubertal male rabbits. *Int J Androl* **15**, 246–254.

Adams D.R. (1986, not seen). The bovine vomeronasal organ. *Arch Histol Jpn* **49**, 211–225.

Adams D.R. (1992). Fine structure of the vomeronasal and septal olfactory epithelia, and of glandular structures. *Micro Res Techn* **23**, 86–97.

Adams D.R. and Wiekamp M.D. (1984). The canine vomeronasal organ. *J Anat* **138**, 771–787.

Adams M., Teeter J.H., Katz Y. and Johnsen P.B. (1987). Sex pheromones of the sea lamprey: steroid studies on urinary products. *J Chem Ecol* **13**, 387–395.

Adams M.G. (1980). Odour-producing organs of mammals. In: *Mammalian Olfaction* (Stoddart D.M., ed.). Academic Press, London, pp. 57–86.

Afshar M., Hubbard R. and Demaille J. (1998). Towards structural models of molecular recognition in olfactory receptors. *Biochimie* **80**, 129–135.

Al-Shawi R., Ghazal P., Clark J. and Bishop J.O. (1989). Intraspecific evolution of a gene family coding for urinary proteins. *J Molec Evol* **29**, 302–313.

Alberts A., Jackintell L. and Phillips J. (1994). Effects of chemical and visual exposure to adults on growth, hormones, and behavior of juvenile Green Iguanas. *Physiol Behav* **55**, 987–992.

Albone E. (1997). Mammalian semiochemistry: chemical signaling between mammals. In: *Handbook of Biosensors and Electronic Noses* (Kress-Rodgers E., ed.). CRC Press, Boca Raton, pp. 503–519.

Allen W.K. and Akeson R. (1985). Identification of an olfactory receptor neuron subclass: cellular and molecular analysis during development. *Dev Biol* **109**, 393–401.

Alonso J., Arevalo R., Garciaojeda E., Porteros A., *et al.* (1995). NADPH-diaphorase active and calbindin d-28k-immunoreactive neurons and fibers in the olfactory-bulb of the hedgehog (*Erinaceus europaeus*). *J Comp Neurol* **351**, 307–327.

Altner H. and Müller W. (1968). Electrophysiological and electron microscopical investigation of the sensory epithelium in the vomeronasal organ in lizards (*Lacerta*). *Z Vergl Physiol* **60**, 151–155.

Altner H., Müller W. and Brachner I. (1970). Ultra-structure of the vomeronasal organ in Reptilia. *Z Zellforsch* **105**, 107–122.

Altieri R. and Müller-Schwarze D. (1980). Seasonal changes in flehmen to constant urine stimuli. *J Chem Ecol* **6**, 905–909.

Alving W.R. and Kardong K.V. (1996). The role of the vomeronasal organ in rattlesnake (*Crotalus viridis oreganus*) predatory behavior. *Brain Behav Evol* **48**, 165–172.

Andren C. (1982). The role of the vomeronasal organs in the reproductive behaviour of the Adder, *Vipera berus*. *Copeia*, 148–157.

Andres K. (1970). Anatomy and ultrastructure of the olfactory bulb in fish, amphibia, reptiles, birds and mammals. In: *Taste and Smell in Vertebrates* (Wolstenholme G. and Knight J., eds.). J&A Churchill, London, pp. 177–193.

Armstrong, J.A., Gamble, H.J. and Goldby, F. (1953). Observations on the olfactory apparatus and telencephalon of *Anolis*, a microsmatic lizard. *J Anat* **87**, 288-307.

Arnautovic I., Abdalla O. and Fahmy M. (1970). Anatomical study of the vomeronasal organ and the nasopalatine duct of the one-humped camel. *Acta Anat* **77**, 144–154.

Arnold S.J. (1977). The evolution of courtship behavior in New World salamanders. In: *The Reproductive Biology of Amphibians* (Taylor D.H. and Guttman S.I., eds.). Plenum, New York, pp. 141–183.

Asano-Miyoshi M., Suda T., Yasuoka A., *et al.* (2000). Random expression of main and vomeronasal olfactory receptor genes in immature and mature olfactory epithelia of *Fugu rubripes*. *J Biochem* **127**, 915–924.

Astic L., Le Pendu J., Mollicone R., Saucier D. and Oriol R. (1989). Cellular expression of H&B antigens in the rat olfactory system during development. *J Comp Neurol* **289**, 386–394.

Aujard F. (1997). Effect of vomeronasal organ removal on male sociosexual responses to females in a prosimian primate (*Microcebus murinus*). *Physiol Behav* **62**, 1003–1008.

Bacchini A., Gatetani E. and Cavaggioni A. (1992). Pheromone binding proteins of the mouse, *Mus musculus*. *Experientia* **48**, 419–421.

Badenhorst A. (1978). The development and the phylogeny of the Organ of Jacobson and the tentacular apparatus of *Ichthyopsis glutinosus* (*Linné*). *Ann Univ Stellenb* **1**(A, 2), 1–26.

Bailey K. (1978). Flehmen in the Ring-tailed Lemur (*Lemur catta*). *Behaviour* **65**, 309–319.

Bakker J. and Baum M. (2000). Neuroendocrine regulation of GnRH release in induced ovulators. *Front Neuroendocrinol* **21**, 220–262.

Bannister L.H. (1968). Fine structure of the sensory endings in the vomeronasal organ of the Slow-Worm *Anguis fragilis*. *Nature* **217**, 275–276.

Bannister L.H. and Dodson H.C. (1992). Endocytic pathways in the olfactory and vomeronasal epithelia of the mouse: ultrastructure and uptake of tracers. *Micros Res Techn* **23**, 128–141.

Barber P.C. and Raisman G. (1971). The projection of the amygdala to the accessory olfactory bulb in the mouse. *Exp Brain Res* **14**, 395–404.

Barber P.C. and Raisman G. (1974). An autoradiographic investigation of the projection of the vomeronasal organ to the accessory olfactory bulb in the mouse. *Brain Res* **81**, 21–30.

Bardach J.D. and Villars T. (1968). The chemical senses of fishes. In: *The Central Nervous System and Fish Behavior* (Ingle D., ed.). Chicago University Press, pp. 48–60.

Bardach J.E. and Todd J.H. (1970). Chemical communication in fish. In: *Advances in Chemoreception* **1**: *Communication by Chemical Signals* (Johnston J.W. Jr., Moulton D.G. and Turk A., eds.). Appleton-Century-Crofts, New York, pp. 10–23.

Barfield R.J. and Thomas D.A. (1986). The role of ultrasonic vocalizations in the regulation of reproduction in rats. *Ann NY Acad Sci* **474**, 33–43.

Bargmann C. (1997). Olfactory receptors, vomeronasal receptors, and the organisation of olfactory information. *Cell* **90**, 585–587.

Barkley M., DeLeon D. and Weste R. (1993). Pheromonal regulation of the mouse estrous cycle by a heterogenotypic male. *J Exp Zool* **265**, 558–566.

Barrette C. (1976). Musculature of facial scent glands in the muntjac. *J Anat* **122**, 16–26.

Barone R. and Lombard M. (1966). Organe de Jacobson, nerf voméro-nasal et nerf terminal du chien. *Bull Soc Sci Vet Med Comp Lyon*, 257–270.

Barrett J., Abbott D. and George L. (1993). Sensory cues and the suppression of reproduction in subordinate female marmoset monkeys, *Callithrix jacchus*. *J Reprod Fertil* **97**, 301–310.

Baxter R.M. (1981). Flehmen in two Southern African shrew species. *Mammalia* **45**, 379–380.

Bean N.J. and Wysocki C.J. (1985). Behavioural effects of removal of the vomeronasal organ in neonatal mice. *Chem Senses* **10**, 421–422.

Bean N.J. and Wysocki C.J. (1987). Effects of vomeronasal organ removal in lactating female mice. *Ann NY Acad Sci* **510**, 169–170.

Bean N.J. and Wysocki C.J. (1989). Vomeronasal organ removal and female mouse aggression: the role of experience. *Physiol Behav* **45**, 875–882.

Beauchamp G., Doty R.L., Moulton D.G. and Mugford R.A. (1976). The pheromone concept in mammalian chemical communication: a critique. In: *Pheromones and Reproduction in Mammals* (Doty R.L., ed.). Academic Press, New York, pp. 143–160.

Beauchamp G.K., Magnus J.G., Shmunes N.T. and Durham T. (1977). Effects of olfactory bulbectomy on social behavior of male guinea pigs (*Cavia porcellus*). *J Comp Physiol Psychol* **91**, 336–346.

Beauchamp G.K., Martin I., Wysocki C.J. and Wellington J.L. (1982). Chemoinvestigatory and sexual behavior of male guinea pigs following vomeronasal organ removal. *Physiol Behav* **29**, 329–336.

Beauchamp G.K., Martin I., Wysocki C.J. and Wellington J.L. (1983). The accessory olfactory system: role in maintenance of chemoinvestigatory behavior. In: *Chemical Signals in Vertebrates* **3** (Müller-Schwarze D. and Silverstein R., eds.). Plenum, New York, pp. 73–86.

Beauchamp G.K., Wysocki C.J. and Wellington J.L. (1985). Extinction of response to urine odor as a consequence of vomeronasal organ removal in male guinea pigs. *Behav Neurosci* **99**, 950–955.

Belcher A.M., Epple G., Greenfield K.L., Richards L.E., *et al.* (1990). Proteins: biologically relevant components of the scent marks of a primate (*Saguinus fusicollis*). *Chem Senses* **15**, 431–446.

Bellairs A.D. and Boyd J. (1950). The lachrymal apparatus in lizards and snakes, II. *Proc Zool Soc Lond* **120**, 269–310.

Bellringer J.F. *et al.* (1980). Involvement of the vomeronasal organ and prolactin in pheromonal induction of delayed implantation in mice. *J Reprod Fertil* **59**, 223–228.

Belluscio L., Gold G., Nemes A. and Axel R. (1998). Mice deficient in Golf are anosmic. *Neuron* **20**, 69–81.

Belluscio L., Koentges G., Axel R. and Dulac C. (1999). A map of pheromone receptor activation in the mammalian brain. *Cell* **97**, 209–220.

Beltramino C. and Taleisnik S. (1983). Release of LH in the female rat by olfactory stimuli. Effect of the removal of the vomeronasal organs or lesioning of the accessory olfactory bulbs. *Neuroendocrinology* **36**, 53–58.

Ben-Arie N., Lancet D., Taylor C., Khen M., *et al.* (1994). Olfactory receptor gene cluster on human Chromsome 17: possible duplication of an ancestral receptor repertoire. *Hum Mol Genet* **3**, 229–235.

Berghard A., Buck L. and Liman E. (1996). Evidence for distinct signaling mechanisms in two mammalian olfactory sense organs. *Proc Natl Acad Sci* **93**, 2365–2369.

Berghard A. and Buck L.B. (1996). Sensory transduction in vomeronasal neurons: evidence for G-alpha-o, G-alpha-i2, and adenylyl cyclase II as major components of a pheromone signaling cascade. *J Neurosci* **16**, 909–918.

Berghard A. and Dryer L. (1998). A novel family of ancient vertebrate odorant receptors. *Neurobiol J* **37**, 383–392.

Bertmar G. (1969). The vertebrate nose, remarks on its structural and functional adaptation and evolution. *Evolution* **23**, 131–152.

Bertmar G. (1981). Evolution of vomeronasal organs in vertebrates. *Evolution* **35**, 359–366.

Beynon R.J., Robertson D., Hubbard S., Gaskell S. and Hurst J. (1999). The role of protein binding in chemical communication, MUPs in the house mouse. In: *Advances in Chemical Signals in Vertebrates* (Johnston R.E., Müller-Schwarze D. and Sorenson P., eds). Kluwer/Plenum, New York, pp. 137–147.

Bhatnagar K., Wible J.R. and Karim K.B. (1996). Development of the vomeronasal organ in *Rousettus leschenaulti*, Megachiroptera, Pteropidae. *J Anat* **188**, 129–136.

Bhatnagar K. and Meisami E. (1998). Vomeronasal organ in bats and primates: extremes of structural variability and its phylogenetic implications. *Microsc Res Tech* **43**, 465–475.

Bhatnagar K. and Reid K.H. (1996). The human vomeronasal organ — I: historical perspectives. A study of Ruysch's (1703) and Jacobson's (1811) reports on the vomeronasal organ, with comparative comments and English translations. *Biomed Res* **7**, 219–229.

Bhatnagar K.P. and Wible J.R. (1994). Observations on the vomeronasal organ of the Colugo, *Cynocephalus* (Mammalia, Dermoptera). *Anat Acta* **151**, 43–48.

Bianchet M., Bains G., Pelosi P., Pevsner J., *et al.* (1996). The three-dimensional structure of bovine odorant binding protein and its mechanism of odour recognition. *Nature Struct Biol* **3**, 934–939.

Birke L.A. and Sadler D. (1987). Differences in maternal behavior of rats and the sociosexual development of the offspring. *Dev Psychobiol* **20**, 85–100.

Black-Cleworth P. and Verberne G. (1975). Scent-marking, dominance and flehmen behavior in domestic rabbits in an artificial laboratory territory. *Chem Senses Flav* **1**, 465–494.

Bland K.P. and Cottrell D.F. (1989). The nervous control of intraluminal pressure in the vomeronasal organ of the domestic ram. *Q J Exp Physiol* **74**, 8134–8142.

Blazquez N., French J., Long S. and Perry G. (1988). A pheromonal function for the perineal skin glands in the cow. *Vet Rec* **123**, 49–50.

Blissitt M.J., Bland K.P. and Cottrell D.F. (1990). Discrimination between the odors of fresh estrous and non-estrous ewe urine by rams. *Appl Anim Behav Sci* **25**, 51–60.

Block M., Volpe L. and Hayes M. (1981). Saliva as a chemical cue in the development of social behavior. *Science* **211**, 1062–1064.

Bocskei Z., Groom C., Flower D. Wright C.E., *et al.* (1992). Pheromone binding to two rodent urinary proteins revealed by X-ray crystallography. *Nature* **360**, 186–188.

Boehm N., Roos J. and Gasser B. (1994). LHRH-expressing cells in the nasal septum of human fetuses. *Dev Br Res* **82**, 175–180.

Boinski S. (1992). Olfactory communication among Costa Rican squirrel monkeys: a field study. *Folia Primatol* **59**, 127–136.

Bojsen-Møller F. (1975). Demonstration of terminalis, olfactory, trigeminal and perivascular nerves in the rat nasal septum. *J Comp Neurol* **159**, 245–256.

Bojsen-Møller F. and Fahrenkrug J. (1971). Nasal swell-bodies and cyclic changes in the air passage of the rat and rabbit nose. *J Anat* **110**, 25–37.

Bons N., Silhol S., Barbie V., Mestre-Frances N. and Albe-Fessard D. (1998). A stereotaxic atlas of the Grey Lesser Mouse Lemur brain (*Microcebus murinus*). *Brain Res Bull* **46**, 1–173.

Booth K. and Katz L. (2000). Role of the vomeronasal organ in neonatal offspring recognition in sheep. *Biol Reprod* **63**, 953–958.

Booth W.D. (1987). Factors affecting the pheromone composition of voided boar saliva. *J Reprod Fertil* **81**, 427–432.

Booth W.D. and White C.A. (1988). The isolation and purification of Pheromaxein in porcine sub-maxillary glands and saliva. *J Endocr* **118**, 47–57.

Bossy J. (1980). Development of olfactory and related structures in human embryos. *Anat Embryol* **161**, 225–236.

Boyd J. (1932). The classification of the upper lip in mammals. *J Anat* **67**, 409–416.

Boyer M., Jemiolo B., Andreolini F., Wiesler D. and Novotny M. (1989). Urinary volatile profiles of Pine Vole, *M. pinetorum* and their endocrine dependancy. *J Chem Ecol* **15**, 649–662.

Brant C.L., Schwab T.M., Vandenbergh J.G., Schaefer R.L., *et al.* (1998). Behavioural suppression of female Pine Voles after replacement of the breeding male. *Anim Behav* **55**, 615–627.

Braun C.B. (1996). The sensory biology of the living jawless fishes: a phylogenetic assessment. *Brain Behav Evol* **48**, 262–276.

Breer H. (1994). Signal recognition and chemo-electrical transduction in olfaction. *Biosensors Bioelectron* **9**, 625–632.

Breipohl W., Bhatnagar K., Blank M. and Mendoza A. (1981). Intra-epithelial blood vessels in the vomeronasal neuroepithelium of the rat. A light and electron microscopic study. *Cell Tiss Res* **215**, 465–473.

Breipohl W., Naguro T. and Walker D.G. (1989). Post-natal development of Masera's organ in rat. *Chem Senses* **14**, 649–662.

Brennan P.A., Kendrick K. and Keverne E.B. (1995). Neurotransmitter release in the accessory olfactory bulb during and after the formation of an olfactory memory in mice. *Neuroscience* **69**, 1075–1086.

Brennan P. and Keverne E.B. (2000). Neural mechanisms of olfactory recognition memory. In: *Brain, Perception, Memory — Advances in Cognitive Neuroscience* (Bolhuis J.J., ed.). Oxford University Press, pp. 93–111.

Brennan P.A. and Keverne E.B. (2002). The vomeronasal organ. In: *Handbook of Olfaction and Gustation*, 2nd ed. (Doty R.L., ed.). M. Dekker Inc., N.Y. & Basel (in press).

Brennan P., Schellinck H. and Keverne E.B. (1999). Patterns of expression of the immediate-early gene egr-1 in the accessory olfactory bulb of female mice exposed to pheromonal constituents of male urine. *Neurosci* **90**, 1463–1470.

Briand L., Huet J., Perez V., Lenoir G., *et al.* (2000). Odorant and pheromone binding by aphrodisin, a hamster aphrodisiac protein. *FEBS Letts* **476**, 179–185.

Broillet M.-C. and Firestein S. (1996). Gaseous 2nd messengers in vertebrate olfaction. *J Neurobiol* **30**, 49–57.

Broman I. (1920). Das *Organon vomero-nasale Jacobsonii* — Ein Wassergeruchsorgan! *Anat Helfe* **58**, 137–191.

Bronson F. (1976). Urine marking in mice: causes and effects. In: *Mammalian Olfaction, Reproductive Processes and Behavior* (Doty R.L., ed.). Academic Press, New York, pp. 119–143.

Bronson F. (1979). The reproductive ecology of the house mouse. *Q Rev Biol* **54**, 265–299.

Bronson F. and Coquelin A. (1980). The modulation of reproduction by priming pheromones in house mice: speculations on adaptive function. In: *Chemical Signals in Vertebrates* **2** (Müller-Schwarze D. and Silverstein R.M., eds.). Plenum, New York, pp. 243–266.

Bronson F. and Marsden H.M. (1964). Male-induced synchrony of estrus in Deermice. *Gen Comp Endocr* **4**, 634–637.

Bronson F. and Maruniak J.A. (1975). Male-induced puberty in female mice: evidence for a synergistic action of social cues. *Biol Reprod* **13**, 94–98.

Bronson F. and Maruniak J.A. (1976). Differential effects of male stimuli on follicle stimulating hormone, luteinising hormone and prolactin secretion in prepubertal female mice. *Endocrinology* **98**, 1101–1108.

Bronson F., Singer A. and Macrides F. (1988). Chemical properties of a female mouse pheromone that stimulates gonadotrophin secretion in males. *Biol Reprod* **38**, 193–199.

Broom R. (1895). *Contributions to the Comparative Anatomy of Jacobson's Organ.* Thesis [M.D., Glasg.], pp. 121.

Broom R. (1896). Comparative anatomy of the Organ of Jacobson in marsupials. *Proc Linn Soc NSW* **21**, 591–623.

Broom R. (1897/8). A contribution to the comparative anatomy of the mammalian Organ of Jacobson. *Trans Roy Soc Edinb* **39**(viii), 231–255.

Broom R. (1898). On the Organ of Jacobson in the Hyrax. *J Anat Physiol* **32**, 709–713.

Brouette-Lahlou I., Godinot F. and Vernet Maury E. (1999). The mother rat's vomeronasal organ is involved in detection of dodecyl propionate, the pup's preputial gland pheromone. *Physiol Behav* **66**, 427–436.

Brouette-Lahlou I., Vernet Maury E. Godinot F., *et al.* (1992). Vomeronasal organ sustains anogenital licking in primiparous rats. In: *Chemical Signals in Vertebrates* **6** (Doty R.L. and Müller-Schwarze D., eds.). Plenum, New York, pp. 551–555.

Brown C. (1968). Additional observations on the function of the naso-labial grooves of Plethodontid salamanders. *Copeia*, 728–731.

Brown G. and Godin J. (1997). Anti-predator responses to conspecific and heterospecific skin extracts by Three-spined Sticklebacks: alarm pheromones revisited. *Behaviour* **134**, 1123–1134.

Brown J.L. and Eklund A. (1994). Kin-recognition and the MHC — an integrative review. *Am Nat* **143**, 435–461.

Bruce H.M. (1959). An exteroceptive block to pregnancy in the mouse. *Nature* **184**, 105.

Bruner A.L. (1914). Jacobson's Organ and the respiratory mechanism of Amphibians. *Morphol Jb* **48**, 157–165.

Brunjes P.C., Jazaeri A. and Sutherland M.J. (1992). Olfactory bulb organization and development in Monodelphis domestica, the Grey Short-tailed Opossum. *J Comp Neurol* **320**, 544–554.

Brunjes P.C. and Kishore R. (1998). Unilateral naris closure and the rat accessory olfactory bulb. *Chem Senses* **23**, 717–720.

Buck L. (1993). Receptor diversity and spatial patterning in the mammalian olfactory system. In: *The Molecular Basis of Smell and Taste Transduction* (Chadwick D., *et al.*, eds). John Wiley, London, Ciba Symposium **179**, pp. 51–67.

Buck L. (2000). Molecular architecture of odor and pheromone sensing in mammals. *Cell* **100**, 611–618.

Buettner J., Glusman G., Ben-Arie N., Ramos, P., *et al.* (1998). Organization and evolution of olfactory receptor genes on human chromosome 11. *Genomics* **53**, 56–68.

Buglass A., Darling F. and Waterhouse J. (1990). Analysis of the anal sac secretion of the Hyaenidae. In: Chemical Signals in Vertebrates **5** (MacDonald D., Müller-Schwarze D. and Natynczuk S.E., eds.). Oxford University Press, Oxford, pp. 65–69.

Buiakova O., Baker H., Scott J., Farbman A., *et al.* (1996). Olfactory marker protein (OMP) gene deletion causes altered physiological activity of olfactory sensory neurons. *Proc Natl Acad Sci* **93**, 9858–9863.

Buiakova O., Krishna N.S., Getchell T.V. and Margolis F. (1994). Human and rodent OMP genes: conservation of structural and regulatory motifs and cellular localization. *Genomics* **20**, 452–462.

Bulfone A., Wang F., Pevner R., Anderson S., *et al.* (1998). An olfactory sensory map develops in the absence of normal projection neurons or GABAergic interneurons. *Neuron* **21**, 1273–1282.

Burger B.V. and Pretorius P.J. (1988). Identification of thiazole derivatives in the preobital gland secretions of the Gray Duiker *Sylvicapra grimnia*, and the Red Duiker *Cephalophus natalensis*. *Z Naturforsch (C)* **43**, 731–736.

Burghardt G. (1980). Behavioral and stimulus correlates of vomeronasal functioning in reptiles: feeding, grouping, sex and tongue use. In: *Chemical Signals Vertebrates and Aquatic Invertebrates* **1** (Müller-Schwarze D. and Silverstein R.M., eds.). Plenum, New York, pp. 275–302.

Burton P.R. (1990). Vomeronasal and olfactory nerves of adult and larval bullfrogs: II. Axon terminations and synaptic contracts in the accessory olfactory bulb. *J Comp Neurol* **292**, 624–637.

Buzzell G.R. (1996). The Harderian gland — perspectives. *Micro Res Tech* **34**, 2–5.

Calof A.L. (1995). Intrinsic and extrinsic factors regulating vertebrate neurogenesis. *Curr Opin Neurobiol* **5**, 19–27.

Calof A.L., Mumm J.S., Rim P.C. and Shou J. (1998). The neuronal stem cell of the olfactory epithelium. *J Neurobiol* **36**, 190–205.

Cao Y., Oh B. and Stryer L. (1998). Cloning and localization of two multigene receptor families in goldfish olfactory epithelium. *Proc Natl Acad Sci* **95**, 11987–11992.

Cappello P., Tarozzo G., Benedetto A. and Fasolo A. (1999). Proliferation and apoptosis in the mouse vomeronasal organ during ontogeny. *Neurosci Lett* **266**, 37–40.

Carmanchahi P.D., Aldana Marcos H.J., Ferrari C.C. and Affanni J.M. (1999). The vomeronasal organ of the South American Armadillo *Chaetophractus villosus* (Xenarthra, Mammalia): anatomy, histology and ultrastructure. *J Anat* **195**, 587–604.

Carson K.A. and Burd G.D. (1980). Localization of acetylcholinesterase in the main and accessory bulbs of the mouse by light and electron microscopic histochemistry. *J Comp Neurol* **191**, 353–371.

Chang Y., Kelliher K. and Baum M.J. (2000). Steroidal modulation of scent investigation and marking behaviors in male and female ferrets (*Mustela putorius furo*). *J Comp Psychol* **114**, 401–407.

Charles-Dominique P. (1977). Urine marking and territoriality in *Galago alleni* (Lorisoidea, Primates): a field study by radio telemetry. *Z Tierpsychol* **43**, 113–138.

Chen D. and Haviland-Jones J. (1999). Rapid mood change and human odors. *Physiol Behav* **68**, 241–250.

Chen W.P., Witkin J.W. and Silverman A.J. (1990). Sexual dimorphism in the synaptic input to gonadotropin-releasing hormone neurons. *Endocrinology* **126**, 695–702.

Cherry J. and Lepri J.J. (1986). Sexual dimorphism and gonadal control of ultrasonic vocalizations in adult pine voles *Microtus pinetorum. Horm Behav* **20**, 34–48.

Chess A., Simon I., Cedar H. and Axel R. (1994). Allelic inactivation regulates olfactory receptor gene-expression. *Cell* **78**, 823–834.

Chipman R., Holt J. and Fox K. (1966). Pregnancy failure in laboratory mice after multiple short-term exposure to strange males. *Nature* **216**, 653.

Chizar D. Walters A., Urbaniak J., Smith H., *et al.* (1999). Discrimination between envenomed and non-envenomed prey by W. Diamond-backed rattlesnakes (*C. atrox*). *Copeia* **3**, 640–648.

Christensen T. and White J. (2000). Olfactory information processing in the brain. In: *The Neurobiology of Taste and Smell*, 2nd ed. (Finger T.E., Silver W.L. and Restrepo D., eds.). Plenum, New York, pp. 201–232.

Chuah M. and Zheng D. (1987). Olfactory marker protein is present in olfactory receptor cells of human fetuses. *Neuroscience* **23**, 363–370.

Chuah M. and Zheng D. (1992). The human primary olfactory pathway: fine structural and cytochemical aspects during development, and in adults. *Micros Res Tech* **23**, 76–85.

Chuah M.I. and Farbman A.I. (1994). Developmental anatomy of the olfactory system. In: *Handbook of Olfaction and Taste*, 1st ed. (Doty R.L., ed.). M. Decker, N.Y., pp. 147–162.

Ciges M., Labella T., Gayoso M. and Sanchez G. (1977). Ultrastructure of the Organ of Jacobson and comparative study of olfactory mucosa. *Acta Oto-Laryngol* **83**, 47–58.

Clancy A.N., Coquelin A., Macrides F., Gorski R. and Noble E. (1984). Sexual behavior and aggression in male mice: involvement of the vomeronasal system. *J Neurosci* **4**, 2222–2229.

Clancy A.N., Schoenfeld T., Forbes W. and Macrides F. (1994). The spatial organization of the peripheral olfactory system of the hamster, II: receptor surfaces and odorant passageways within the nasal cavity. *Brain Res Bull* **34**, 211–241.

Clancy A.N., Singer A., Macrides F., Bronson F. and Agosta W. (1988). Experiential and endocrine dependance of GnTH responses in male mice to conspecific urine. *Biol Reprod* **38**, 183–191.

Clancy A.N., Macrides F., Singer A. and Agosta W.C. (1984). Male hamster copulatory responses to a high molecular weight fraction of vaginal discharge: effects of vomeronasal organ removal. *Physiol Behav* **33**, 653–660.

Clarke D. (1981). The tongue-vomeronasal transfer mechanism of snakes. *Micron* **12**, 299–300.

Clarris H. and Key B. (2001). Expression of glycoproteins in the vomero-nasal organ reveals a novel spatiotemporal pattern of sensory neurone maturation. *J Neurobiol* **46**, 113–125.

Clissold P. and Bishop J. (1982). Variation in mouse MUP genes and gene products within inbred lines. *Gene* **18**, 211–220.

Coates E.L. and Ballam G.O. (1989). Breathing and upper airway CO_2 in reptiles: role of the nasal and vomeronasal systems. *Am J Physiol* **256**, 91–97.

Cocke R., Moynihan J.A., Cohen N., Grota L.J., *et al.* (1993). Exposure to conspecific alarm signals alters immune responses in BALB-c mice. *Brain Behav Immunol* **7**, 36–46.

Cohen-Tannoudji J., Lavenet C., Locatelli A., Teillet Y. and Signoret J.P. (1989). Non involvement of the accessory olfactory system in the LH reponse of anestrous ewes to male odor. *J Reprod Fertil* **86**, 135–144.

Collado P. and Segovia S. (1992). Female's DHT controls sex differences in the rat bed nucleus of the accessory olfactory tract. *Neuroreport* **3**, 327–329.

Comfort A. (1971). Likelyhood of human pheromomes. *Nature* **230**, 432–443.

Conely L. and Bell R.W. (1978). Neonatal ultrasounds elicited by odor cues. *Dev Psychobiol* **11**, 193–197.

Constanzo R. (1991). Regeneration of olfactory receptor cells. In: *Regeneration of Vertebrate Sensory Cells* (Bock G.R., ed.). Wiley, New York, pp. 233–248.

Converse L., Carlson A.A., Snowdon C.T., Ziegler T. and Snowdon C.T. (1995). Communication of ovulatory state to mates by female Pygmy Marmosets, *Cebuella pygmaea. Anim Behav* **49**, 615–621.

Cooke B., Hegstrom C.D., Villeneuve L.S. and Breedlove S.M. (1998). Sexual differentiation of the vertebrate brain: principles and mechanisms. *Front Neuro-endocrinol* **19**, 323–362.

Coon H., Curcio F., Sakaguchi K., Brandi M.L. and Swerdlow R.D. (1989). Cell cultures of neuroblasts from rat olfactory epithelium that show odourant responses. *Proc Natl Acad Sci* **86**, 1703–1707.

Cooper A. (1974). Effects of accessory olfactory bulb lesions on the sex behavior of male mice. *Bull Psychonom Sci* **1**, 419–420.

Cooper J. and Bhatnagar K.P. (1976). Comparative anatomy of the vomeronasal organ complex in bats. *J Anat* **122**, 571–601.

Cooper W.E. Jr. (1994). Chemical discrimination by tongue-flicking in lizards: a review. *J Chem Ecol* **20**, 439–487.

Cooper W.E. Jr. (1995). Effects of estrogen and male head coloration on chemosensory investigation of female cloacal pheromones by male broad-headed skinks (*Eumeces laticeps*). *Physiol Behav* **58**, 1221–1225.

Cooper W.E. Jr. (1997). Correlated evolution of prey chemical discrimination with foraging, lingual morphology and vomeronasal chemoreceptor abundance in lizards. *Behav Ecol Sociobiol* **41**, 257–265.

Cooper W.J., van Wyk J. and Mouton P. (1999). Discrimination between self-produced pheromones and those produced by individuals of the same sex in the lizard *Cordylus cordylus*. *J Chem Ecol* **25**, 197–208.

Coopersmith C. and Lenington S. (1998). Pregnancy block in house mice (*Mus domesticus*) as a function of t-complex genotype: examination of the mate choice and male infanticide hypotheses. *J Comp Psychol* **112**, 82–91.

Coppola D.M., Budde J. and Millar L.C. (1993). The vomeronasal duct has a protracted postnatal development in mouse. *J Morphol* **218**, 59–64.

Coppola D.M. and Millar L.C. (1994). Stimulus access to the accessory olfactory system in the prenatal and perinatal rat. *Neuroscience* **60**, 463–468.

Coppola D.M. and O'Connell R.J. (1989). Stimulus access to olfactory and vomeronasal receptors *in utero*. *Neuroscience Lett* **106**, 241–248.

Coppola D.M. and Vandenbergh J.G. (1985). Effect of density, duration of grouping and age of urine stimulus on the puberty delay pheromone in female mice. *J Reprod Fertil* **73**, 517–522.

Coppola D.M. and Vandenbergh J.G. (1987). Induction of a puberty-regulating chemosignal in wild house mouse population. *J Mammal* **68**, 86–91.

Cornwall-Jones C.A. (1988). DSP4 a noradrenergic neurotoxin, impairs male rats attraction to conspecific odors. *Behav Neural Biol* **50**, 1–15.

Cornwell-Jones C.A. and Azar L.M. (1982). Olfactory development in gerbil pups. *Dev Psychobiol* **15**, 131–133.

Corotto F., Henegar J. and Maruniak J. (1994). Odor deprivation leads to reduced neurogenesis and reduced neuronal survival in the olfactory bulb of the adult mouse. *Neuroscience* **61**, 739–744.

Coulson G. and Croft D. (1981). Flehmen in Kangaroos. *Aust Mammal* **4**, 139–140.

Cowles R.B. and Phelan R.L. (1958). Olfaction in Rattlesnakes. *Copeia* **2**, 77–83.

Crespo C., Brinon J.G., Porteros A., Arevalo R., *et al.* (1999). Distribution of acetylcholinesterase and choline acetyltransferase in the main and accessory olfactory bulbs of the hedgehog (*Erinaceus europaeus*). *J Comp Neurol* **403**, 53–67.

Crewe R.M., Burger B.V., Le Roux M. and Katsir Z. (1979). Chemical constituents of the chest gland secretion of the Thick-tailed galago (*Galago crassicaudatus*). *J Chem Ecol* **5**, 861–868.

Crowell-Davis S. and Houpt K. (1985). The ontogeny of flehmen in horses. *Anim Behav* **33**, 739–745.

Cummings D. and Brunjes P. (1995). Migrating luteinizing-hormone-releasing hormone (LHRH) neurons and processes are associated with a substrate that expresses S100. *Dev Brain Res* **88**, 148–157.

Cummings D., Knsab B. and Brunjes P. (1997). Effects of unilateral olfactory deprivation in the developing opossum *Monodelphis domestica. J Neurobiol* **33**, 429–438.

Cutler W.B. (1987). Female essence pheromones increases sexual behavior in young women. *Neuroendocrinol Lett* **9**, 199.

Cuvier G. (1812). Rapport fait à l'Institut, sur un Mémoire de M. Jacobson. *Ann Mus d'Hist Nat Paris* **18**, 412–424.

Dagg H. and Taub A. (1970). Flehmen. *Mammalia* **34**, 686–695.

Daikoku S. (1999). The olfactory origin of luteinizing hormone-releasing hormone (LHRH) neurons. A new era in reproduction physiology. *Arch Histol Cytol* **62**, 107–117.

Daikoku S., Koide I., Chikamoriaoyama M. and Shimomura Y. (1993). Migration of LHRH neurons derived from the olfactory placode in rats. *Arch Histol Cytol* **56**, 353–370.

Dantzer R. (1998). Vasopressin, gonadal steroids and social recognition. *Prog Brain Res* **119**, 409–441.

Darney K.J.J., Goldman J.M. and Vandenbergh J.G. (1992). Neuroendocrine responses to social regulation of puberty in the female house mouse. *Neuroendocrinol* **55**, 434–443.

D'Angelo W., Macrides F. and Bartke A. (1974). Effects of exposure to vaginal odor and receptive females on plasma testosterone in the male hamster. *Neuroendocrinol* **15**, 355–364.

Davis B.J., Macrides F., Youngs W.M., Schneider S.P., *et al.* (1978). Efferents and centrifugal afferents of the main and accessory olfactory bulbs in the hamster. *Brain Res Bull* **3**, 59–72.

Dawley E.M. (1998). Species, sex and seasonal differences in vomeronasal organ size. *Micros Res Tech* **41**, 506–518.

Dawley E.M. and Bass A.H. (1988). Organization of the vomeronasal organ in a plethodontid salamander. *J Morphol* **198**, 243–255.

de Boer J. (1977). The age of olfactory cues functioning in chemo-communication among domestic male cats. *Behav Proc* **2**, 209–225.

de Catanzaro D., Muir C., Sullivan C. and Boissy A. (1999). Pheromones and novel male-induced pregnancy disruptions in mice: exposure to conspecifics is necessary for urine alone to induce an effect. *Physiol Behav* **66**, 153–157.

de Fanis E. and Jones G. (1995). The role of odour in the discrimination of conspecifics by Pipistrelle Bats. *Anim Behav* **49**, 835–839.

de Leon D.D. and Barkley M.S. (1987). Male and female genotype mediates pheromonal regulation of the mouse estrous cycle. *Biol Reprod* **37**, 1066–1074.

De Olmos J.S., Hardy H. and Heimer L. (1978). The afferent connections in the main and accessory olfactory bulb formations in the rat: an experimental HRP study. *J Comp Neurol* **181**, 213–244.

De Olmos J.S., *et al.* (1980). The afferent projections of the main and accessory bulb in the Armadillo. *Anat Res* **196A**, 45.

Dehasse J. (1997). Feline urine spraying. *Appl Anim Behav* **52**, 365–371.

Del Cerro M., Chirino R., Mayer A.D. and Rosenblatt J.S. (1995). Sex-differences in sensitisation latencies in maternal behavior induced in male and female rats after vomeronasal organ removal. *Dev Psychobiol* **28**, 183 abs.

Del Cerro M. and Cruz R. (1998). Role of the vomeronasal input in maternal behavior. *Psychoneuroendocrinol* **23**, 905–926.

Delheusy V., Toubeau G. and Bels V. (1994). Tongue structure and function in *Oplurus cuvieri* (Iguanidae). *Anat Rec* **238**, 263–276.

Del Punta K., Rothman A., Rodriguez I. and Mombaerts P. (2000). Sequence diversity and genomic organization of vomeronasal receptor genes in the mouse. *Genome Res* **10**, 1958–1967.

Delville Y., De Vries G.J. and Ferris C.F. (2000). Neural connections of the anterior hypothalamus and agonistic behavior in golden hamsters. *Brain Behav Evol* **55**, 53–76.

Demas G.E., Williams J.M. and Nelson R.J. (1997). Amygdala but not hippocampal lesions impair olfactory memory for mate in prairie voles (*Microtus ochrogaster*). *Am J Physiol* **273**, 1683–1689.

Deniker J. (1885). Recherches Anatomique et Embryologiques sur les singes anthropoides, Foetus de Gorille et de Gibbon. *Arch Zool Exp Gen* **2e**, **III**, bis: Suppl.

Derivot J. (1984). Functional anatomy of the peripheral olfactory system of the African lungfish *Protopterus annectens*, Owen: macroscopic, microscopic and morphometric aspects. *Am J Anat* **169**, 177–192.

Devitsina G.V. and Cherova L. (1992). The trigeminal nerve system and its interaction with olfactory and taste system in fishes. In: *Chemical Signals in Vertebrates* **6** (Doty R.L. and Müller-Schwarze D., eds.). Plenum, New York, pp. 85–88.

Dieterlen F. (1959). Das verhalten des Syrischen Goldhamsters (*Mesocricetus auratus*). *Z Tierpsychol* **16**, 47–103.

Dionne V.E. (1994). Transduction diversity in olfaction. *J Exp Biol* **194**, 1–21.

Dixson A.F. (1983). Observations on the evolution and behavioural significance of sexual skin in female primates. *Adv Stud Behav* **13**, 63–106.

Dizinno G., Whitney G. and Nyby J. (1978). Ultrasonic vocalisation by male mice (*Mus musculus*) to female sex pheromone: experiential determinants. *Behav Biol* **22**, 104–113.

Dorries K.M., Adkins-Regan E. and Halpern B.P. (1995). Olfactory sensitivity to the pheromone, androstenone, is sexually dimorphic in the pig. *Physiol Behav* **57**, 255–259.

Dorries K.M., Adkins-Regan E. and Halpern B.P. (1997). Olfactory sensitivity and behavioral responses to the pheromone androstenone are not mediated by the vomeronasal organ in domestic pigs. *Brain Behav Evol* **49**, 53–62.

Døving K.B., Trotier D., Rosin J.-F. and Holey A. (1993). Functional architecture of the vomeronasal organ of the frog (*Rana*). *Acta. Zool Stockh* **74**, 173–180.

Døving K.B. and Trotier D. (1998). Structure and function of the vomeronasal organ. *J Exp Biol* **201**, 2913–2925.

Drickamer L.C. (1989). Patterns of deposition of urine containing chemosignals that affect puberty and reproduction in house mice. *J Chem Ecol* **15**, 1407–1421.

Drickamer L.C. and Brown P.L. (1998). Age-related changes in odor preferences by house mice living in semi-natural enclosures. *J Chem Ecol* **24**, 1745–1756.

Dryer L. and Berghard A. (1999). Odourant receptors: a plethora of G-protein-coupled receptors. *Trends Pharmacol Sci* **20**, 413–419.

Dryer L. and Graziadei P. (1993). A pilot study on morphological compartmentalization and heterogeneity in the elasmobranch olfactory bulb. *Anat Embryol (Berl)* **188**, 41–51.

Dudley C.A. and Moss R.L. (1994). Lesions of the accessory olfactory bulb decrease lordotic responsiveness and reduce mating-induced c-fos expression in the accessory olfactory system. *Brain Res* **642**, 29–37.

Dudley C.A. and Moss R.L. (1996). Signal processing in the vomeronasal system, modulation of sexual behavior in the female rat. *Crit Revs Neurobiol* **10**, 265–290.

Dudley C.A. and Moss R.L. (1999). Activation of an anatomically distinct subpopulation of accessory olfactory bulb neurons by chemosensory stimulation. *Neuroscience* **91**, 1549–1556.

Dugmore S.J. (1984). *Chemical Signalling Mechanisms in Juvenile and Adult Lemur Catta*, L. Ph.D. Thesis, Glasgow Caledonian University.

Dulac C. (2000). Sensory coding of pheromone signals in mammals. *Curr Opin Neurobiol* **10**, 511–518.

Dulac C. and Axel R. (1995). A novel family of genes encoding putative pheromone receptors in mammals. *Cell* **83**, 195–206.

Dulac C. and Axel R. (1998). Expression of candidate pheromone receptor genes in vomeronasal neurons. *Chem Senses* **23**, 467–475.

Dulka J. (1993). Sex pheromone systems in goldfish: comparable to vomeronasal systems in Tetrapods? *Brain Behav Evol.* **42**, 265–280.

Dunbar I.F. and Beach F.A. (1980). Developmental and activational effects of sex hormones on the attractiveness of dog urine. *Physiol Behav* **24**, 201–204.

Duvall D. (1981). Western fence lizard (*Sceloporus occidentalis*) chemical signals, II: a replication with naturally breeding adults and a test of the Cowles and Phelan hypothesis of rattlesnake olfaction. *J Exp Zool* **218**, 351–361.

Duvall D. (1986). A new question of pheromones: Aspects of possible chemical signalling and reception in the mammal-like reptiles. In: *Ecology and Biology of Mammal-like Reptiles* (Hotton N., *et al.*, eds.). Smithsonian Institution, Washington D.C., pp. 219–238.

Duvall D., King M. and Graves B. (1983). Fossil and comparative evidence for possible chemical signaling in the mammal-like reptiles. In: *Chemical Signals in Vertebrates* **3** (Müller-Schwarze D. and Silverstein R., eds.). Plenum, New York, pp. 25–44.

Eales N.B. (1926). The anatomy of the head of a foetal African elephant. *Phil Trans Roy Soc Edin* **LIV**(III), 491–547.

Ebrahimi F.A. and Chess A. (2000). Olfactory neurons are interdependent in maintaining axonal projections. *Curr Biol* **10**, 219–222.

Ebrahimi F.A., Edmondson J., Rothstein R. and Chess A. (2000). YAC transgene-mediated olfactory receptor gene choice. *Dev Dyn* **217**, 225–231.

Eccles R. (1982). Autonomic innervation of the vomeronasal organ of the cat. *Physiol Behav* **28**, 1011–1015.

Eccles R. (1983). Sympathetic control of nasal erectile tissue. *Eur J Respir Dis* Suppl. **128**, 150–154.

Eggert F., Luszyk D., Haberkorn K., Wobst B., *et al.* (1998/9). The major histocompatibility complex and the chemosensory signalling of individuality in humans. *Genetica* **104**, 265–273.

Eisenberg J.F., MacKay G. and Jainudeen J.R. (1971). Sex behavior of Asiatic elephant (*Elephas maximus*). *Behaviour* **38**, 193–225.

Eisthen H., Wysocki C.J. and Beauchamp G. (1987). Behavioral responses of male guinea pigs to conspecific signals following neonatal vomeronasal organ removal. *Physiol Behav* **41**, 445–449.

Eisthen H.L. (1992). Phylogeny of the vomeronasal system and of receptor cell-types in the olfactory and vomeronasal epithelia of vertebrates. *Microsc Res Tech* **23**, 1–21.

Eisthen H.L. (1997). Evolution of vertebrate olfactory systems. *Brain Behav Evol* **50**, 222–233.

Eisthen H.L. (2000a). Presence of the vomeronasal system in aquatic salamanders. *Philos Trans Roy Soc Lond B* **355**, 1209–1213.

Eisthen H.L. (2000b). Neuromodulatory effects of gonadotropin releasing hormone on olfactory receptor neurons (*Necturus maculosus*). *Neuroscience* **20**, 3947–3955.

Eisthen H.L., Sengelaub D., Schroeder D. and Alberts J. (1994). Anatomy and forebrain projections of the olfactory and vomeronasal organs in Axolotls (*Ambystoma mexicanum*). *Brain Behav Evol* **44**, 108–124.

Eisthen H.L., Wysocki C.J. and Beauchamp G.K. (1987). Behavioral responses of male guinea pigs to conspecific chemical signals following neonatal vomeronasal organ removal. *Physiol Behav* **41**, 445–449.

el Amraoui A. and Dubois P. (1993). Experimental evidence for early commitment of GnRH neurons. *Neuroendocrinol* **57**, 991–1002.

Epple G. (1978). Studies on the nature of chemical signals in scent marks and urine of *Saguinus fuscicollis* (Callitrichidae, Primates). *J Chem Ecol* **4**, 383–394.

Epple G. (1986). Communication by chemical signals. In: *Comparative Primate Biology*, v. **2.b** (Mitchell G., ed.). Liss Inc., N.Y., pp. 531–580.

Epple G., Alveario M.C., Golob N.F. and Smith A.B. (1980). Stability and attractiveness related to age of scent marks of Saddle-Back Tamarins (*Saguinus fuscicollis*). *J Chem Ecol* **6**, 735–748.

Estes J.A. (report of Riedman M. and Deutsch C.) (1989). Adaptations for aquatic living by carnivores. In: *Carnivore Behavior, Ecology and Evolution* (Gittleman J.L., ed.). Chapman & Hall, London, pp. 248–249.

Estes R.D. (1972). The role of the vomeronasal organ in mammalian reproduction. *Mammalia* **36**, 315–341.

Evans C.S. (1980). Diosmic responses to scent-signals in *Lemur catta*. In: *Chemical Signals Vertebrates and Aquatic Invertebrates* **2** (Muller-Schwarze D. and Silverstein R.M., eds.). Plenum, New York, pp. 417–420.

Evans C.S. (1984). On the structure and function of accessory chemoreceptive organs. *Acta Zool Fennica* **171**, 57–62.

Evans C.S. (1998). Vomeronasal-like structures in the Great Apes. Paper presented at *17th Congress of International Primatological Society* (Ramasominana B., ed.). Antananarivo, abs. 290.

Evans C.S. and Goy R.W. (1968). Social behaviour and reproductive cycles in captive Ring-tailed lemurs, *Lemur catta*. *J Zool* **156**, 171–197.

Evans C.S. and Grigorieva E.V. (1995). Morphology of the vomeronasal organ in two South American primates, Red-Bellied Tamarin (*Saguinus labiatus*) and Dwarf Marmoset (*Cebuella pygmaea*): histology and lectin histochemistry. *Adv Biosciences* **93**, 31–42.

Evans C.S. and Grigorieva E.V. (2002) Vomeronasal neuroepithelium in the mid-term human foetus: lectin histochemistry. Paper presented at ECRO Congress, Erlangen. *Chem Senses* **27** (in press).

Evans C.S. and Schilling A. (1995). The accessory (vomeronasal) chemoreceptor system in some Prosimians. In: *Creatures of the Dark: The Nocturnal Prosimians* (Altermann L., Doyle G. and Izard M.K., eds.). Plenum, New York, pp. 393–412.

Fabre-Nys C., Ohkura S. and Kendrick K.M. (1997). Male faeces and odors evoke differential patterns of neurochemical release in the MBH of ewes during estrus: an insight into sexual motivation? *Europ J Neuroscience* **9**, 1666–1677.

Fadem B.H. (1989). The effects of pheromonal stimuli on estrus and peripheral plasma estradiol in female gray short-tailed opossums. *Biol Reprod* **41**, 213–217.

Farbman A. (1991). Developmental neurobiology of the olfactory system. In: *Smell and Taste in Health and Disease* (Getchell T.V., ed.). Raven Press, N.Y., pp. 19–33.

Farbman A.I., Buchholz J., Suzuki Y., Coines A. and Speert D. (1999). A molecular basis of cell death in olfactory epithelium. *J Comp Neurol* **414**, 306–314.

Farbman A.I. and Squnto L.M. (1985). Early development of olfactory receptor cell axons. *Brain Res* **351**, 205–213.

Faulkes C.G. and Abbott D.H. (1993). Evidence that primer pheromones do not cause social suppression of reproduction in male and female Naked Mole Rats (*H. glaber*). *J Reprod Fertil* **99**, 225–230.

Feldhoff R.C., Rollman S.M. and Houck L.D. (1999). Chemical analysis of courtship pheromones in a Plethodontid Salamander. In: *Advances in Chemical Signals in Vertebrates* (Johnston R.E., Müller-Schwarze D. and Sorenson P., eds.). Kluwer, New York, pp. 117–126.

Felix B., Leger M. and Albe-Fessard D. (1999). Stereotaxic atlas of the pig brain. *Brain Res Bull* **49**, 1–136.

Feng W.-H. Kauer J.S., Adelman L. and Talamo B.R. (1997). A new structure — the "olfactory pit" in human olfactory mucosa. *J Comp Neurol* **37**, 443–453.

Ferkin M., Sorokin E. and Johnston R.E. (1997). Effect of prolactin on the attractiveness of male odors to females in Meadow Voles: independance and additive effects with testosterone. *Horm Behav* **31**, 55–63.

Ferkin M.H. and Johnston R.E. (1995). Meadow voles, *Microtus pennsylvanicus*, use multiple sources of scent for sex recognition. *Anim Behav* **49**, 37–44.

Fernandez-Fewell G.D. and Meredith M. (1995). Facilitation of mating behavior in male hamsters LHRH and AcLHRH5-10: interaction with the vomeronasal system. *Physiol Behav* **57**, 213–224.

Ferron J. (1973). Morphologie comparée de l'organe de l'odorat chez quelques mammifères carnivores. *Nat Canad* **100**, 525–541.

Ferstl R., Eggert F., Westphal E., Zavazava N., *et al.* (1992). MHC-related odors in humans. In: *Chemical Signals in Vertebrates* **6** (Doty R.L. and Müller-Schwarze D., eds.). Plenum, New York, pp. 205–212.

Finlay B.L. and Darlington R.B. (1995). Linked regularities in the development and evolution of mammalian brains. *Science* **268**, 1578–1583.

Firestein S. (1996). Scentsational ion channels. *Neuron* **17**, 803–806.

Fischer R. and Brown P. (1993). Vaginal secretions increase the likelihood of inter-male aggression in Syrian hamsters. *Physiol Behav* **54**, 213–214.

Fishelson L. and Baranes A. (1997). Ontogenesis and cytomorphology of the nasal olfactory organs in the Oman shark, *Iago omanensis* (Triakidae), in the Gulf of Aqaba. *Anat Rec* **249**, 409–421.

Fleming A., Vaccarino F., Tambosso L. and Chee P. (1979). Vomeronasal and olfactory system modulation of maternal behaviour in the rat. *Science* **203**, 372–374.

Fletcher T. (1978). The induction of male sexual behavior in Red Deer by the administration of testosterone to hinds, and Estradiol-17β to stags. *Horm Behav* **11**, 74–88.

Flood P.F. (1985). Mammalian chemosignal sources. In: *Social Odours in Mammals* **1** (Brown R.E. and Macdonald D.W., eds.). Clarendon Press, Oxford, pp. 1–32.

Floody O. and Comerci J. (1987). Hormonal control of sex-differences in UHF production by hamsters. *Horm Behav* **2**, 17–35.

Flower D.R. (1995). Multiple molecular recognition properties of the lipocalin protein family. *J Mol Recog* **8**, 185–195.

Flower D.R. (1996). The lipocalin protein family: structure and function. *Biochem J* **318**, 1–14.

Frahm H.D. (1981). Volumetric comparison of the accessory olfactory bulb in bats. *Acta Anat* **109**, 173–183.

Frahm H.D. and Bhatnagar K. (1980). Comparative morphology of the accessory olfactory bulb in bats. *J Anat* **130**, 349–366.

Frahm H.D., Stephan H. and Baron G. (1984). Comparison of accessory olfactory bulb volumes in the Common Tree Shrew (*Tupaia glis*). *Acta Anat* **119**, 129–135.

Franceschini V., Lazzari M. and Ciani F. (1996). Identification of surface glycoconjugates in the olfactory system of turtle. *Brain Res* **725**, 81–87.

Franceschini V., Sbarbati A. and Zancanaro C. (1991). The vomeronasal organ in the frog, *Rana esculenta* — an electron-microscopy study. *J Submicros Cytol Pathol* **23**, 221–231.

Frankiewicz J. and Marchlewska-Koj A. (1985). Effect of conspecifics on sexual maturation in female European Pine Voles (*Pitymys subterraneus*). *J Reprod Fertil* **74**, 153–156.

Freitag J., Ludwig G., Andreini I., Rossler P., *et al.* (1998). Olfactory receptors in aquatic and terrestrial vertebrates. *J Comp Physiol [A]* **183**, 635–650.

Freitag J., Beck A., Ludwig G., von Buchholtz L. and Breer H. (1999). On the origin of the olfactory receptor family: receptor genes of the jawless fish (*Lampetra fluviatilis*). *Gene* **226**, 165–174.

Frets G.P. (1912). Beitrage zur vergleichenden Anatomie und Embryologie der Nase der Primaten — I. *Morphol Jb* **12**, 409–465.

Friedle R. and Fischer R. (1984). Discrimination of salivary olfactants by male hamsters. *Psychol Rep* **55**, 67–70.

Fuchs E., Rosenbusch J. and Anzenberger G. (1991). Urinary protein pattern reflects social rank in male Common Marmosets. *Folia Primatol (Basel)* **57**, 177–180.

Gaafar H.A., Tantawy A., Hamza M. and Shaaban M. (1998). The effect of ammonia on olfactory epithelium and vomeronasal organ neuroepithelium of rabbits: a histological and histochemical study. *J Otorhinolaryngol* **60**, 88–89.

Gaafar H.A., Tantawy A., Melis A., Hennawy D.M. and Shehata H. (1998). The vomeronasal (Jacobson's) organ in adult humans: frequency of occurrence and enzymatic study. *Acta Otolaryngol* **118**, 409–412.

Gabe M. and Saint Girons H. (1976). Contribution à la morphologie comparée des fosses nasales et de leurs annexes chez les Lépidosauriens. *Mem Mus Nat d'hist Nat Paris (A)* **98**, 1–78.

Gabriele O.F. (1967). Persistent vomeronasal organ. *Am J Roentgenol* **99**, 697–699.

Gangrade B. and Dominic C. (1985). Evaluation of the involvement of the vomeronasal organ in the pheromonal influences on the estrous cycle of the laboratory mouse. *Anim Reprod Sci* **8**, 181–185.

Garcia-Velasco J. and Mondragon M.Z. (1991). Incidence of vomeronasal organ in 1000 humans and its possible clinical significance. *J Steroid Biochem Mol Biol* **39**(4b), 561–564.

Garrosa M., Gayoso M.J. and Esteban F.J. (1998). Prenatal development of the mammalian vomeronasal organ. *Microsc Res Tech* **41**, 456–470.

Garrosa M., Szabo K. and Mendoza A.S. (1992). Developmental stages of the vomeronasal epithelium in rat: a light and electron-microscope study. *J Hirnforsch* **33**, 123–132.

Gaupp E., ed. (1904). *Anatomie des Frosches*. Ecker & Wiersheim, Berlin, p. 622.

Geyer L., *et al.* (1978). Effect of ultra high frequency and male urine on female rat readiness to mate. *J Comp Physiol Psychol* **92**, 457–462.

Gheri G., Bryk S. and Balboni G. (1992). Identification of sugar residues in human fetal olfactory epithelium using lectin histochemistry. *Acta Anat* **145**, 167–174.

Giacobini P., Benedetto A., Tirindelli R. and Fasolo A. (2000). Proliferation and migration of receptor neurons in the vomeronasal organ of the adult mouse. *Dev Brain Res* **123**, 33–40.

Giannetti N., Saucier D. and Astic L. (1992). Organization of the septal organ projection to the main olfactory bulb in adult and newborn rats. *J Comp Neurol* **323**, 288–298.

Giannetti N., Saucier D. and Astic L. (1995). Analysis of the possible alerting function of the septal organ in rats — a lesional and behavioral study. *Physiol Behav* **58**, 837–845.

Giannetti N., Pellier V., Oestreicher A.B. and Astic L. (1995). Immunocytochemical study of the differentiation process of the Septal Organ of Masera in developing rats. *Dev Brain Res* **84**, 287–293.

Gilad Y., Segre D., Skorecki K., Nachman M., *et al.* (2000). Dichotomy of single-nucleotide polymorphism haplotypes in olfactory receptor genes and pseudogenes. *Nature Genet* **26**, 221–224.

Gilbertson T., Damak S. and Margolskee R.F. (2000). Molecular physiology of taste transduction. *Curr Opin Neurobiol* **10**, 519–527.

Gilder P.M. and Slater P.J. (1978). Interest of mice in conspecific male odours is influenced by degree of kinship. *Nature* **274**, 364–365.

Gillingham J. and Clark D. (1981). Snake tongue-flicking: transfer mechanisms to Jacobson's Organ. *Can J Zool* **59**, 1651–157.

Glander K.E., Wright P.C., Seigler D.S., Randrianasolo V., *et al.* (1989). Consumption of cyanogenic bamboo by a newly discovered species of Bamboo Lemur. *Am J Primatol* **19**, 119–124.

Glusman G., Bahar A., Sharon D., Pilpel Y., *et al.* (2000). The olfactory receptor gene superfamily: data mining, classification, and nomenclature. *Mamm Genome* **11**, 1016–1023.

Gogos J.A., Osborne J., Nemes A., Mendelsohn M. and Axel R. (2000). Genetic ablation and restoration of the olfactory topographic map. *Cell* **103**, 609–620.

Goodrich B.S. and Myktowycz R. (1972). Individual and sex-differences in chemical composition of pheromone-like substances from skin glands of rabbit. *J Mammal* **53**, 540–548.

Goodwin T.E., Rasmussen L.E.L., Guinn A.C., McKelvey S.S., *et al.* (1999). African elephant sesquiterpenes.[(E)-2, 3-dihydrofarnesol]. *J Nat Prod* **62**, 1570–1572.

Gorman M.L. (1990). Scent marking stratgies in mammals. *Rev Suisse Zool* **97**, 3–30.

Gouda M., Matsutani S., Senba E. and Tohyama M. (1990). Peptidergic granule cell populations in the rat main and accessory olfactory bulb. *Brain Res* **512**, 339–342.

Gove D. (1979). A comparative study of snake and lizard tongue-flicking, with an evolutionary hypothesis. *Z Tierpsychol* **51**, 58–76.

Gower D. and Booth W.D. (1986). Salivary pheromones in the pig and human in relation to sexual status and age. In: *Ontogeny of Olfaction* (Breipohl W., ed.). Springer, Berlin, pp. 255–264.

Graham A. and Begbie J. (2000). Organisation of neurogenic placodes. *Trends Neurosci* **23**, 313–316.

Grau G. (1976). Olfaction and reproduction in Ungulates. In: *Mammalian Olfaction, Reproduction and Behavior* (Doty R.L., ed.). Academic Press, New York, pp. 219–242.

Gravelle K. and Simon C.A. (1980). Field observations on the use of the tongue-Jacobson Organ system in two Iguanid lizards *Scleporus jarrovi*, and *Anolis tritinatis*. *Copeia*, 356–362.

Graves B.M. (1993). Chemical delivery to the vomeronasal organs and functional domain of Squamate chemoreception. *Brain Behav Evol* **41**, 198–202.

Graves B.M. and Duvall D. (1985). Mouth — gaping and head — shaking by Prairie Rattlesnakes are associated with vomeronasal organ olfaction. *Copeia*, 496–497.

Graves B.M. and Duvall D. (1988). Evidence of an alarm pheromone from the cloacal secs of prairie rattlesnakes. *Southwest Nat* **33**, 339–346.

Graves B.M. and Halpern M. (1989). Chemical access to the vomeronasal organ of the lizard *Chacides ocellatus*. *J Exp Zool* **249**, 150–157.

Graves B.M. and Halpern M. (1990). Roles of vomeronasal organ chemoreception in tongue flicking, exploratory and feeding behaviour of the lizard *Chacides ocellatus*. *Anim Behav* **39**, 692–698.

Graves B.M., Halpern M. and Friesen J.L. (1991). Snake aggregation pheromones: source and chemosensory mediation in Western Ribbon snakes (*Thamnophis proximus*). *J Comp Psychol* **105**, 140–144.

Graves R.M., Halpern M. and Gillingham J.C. (1993). Effects of vomeronasal system deafferentation on home range use in a natural population of Eastern Garter snakes, *Thamnophis sirtalis*. *Anim Behav* **45**, 307–311.

Graziadei P.P. (1970). Vomeronasal receptors in turtles. *Z Zellforsch* **105**, 498–514.

Graziadei P.P. (1990). Olfactory development. In: *Development of Sensory Systems in Mammals* (Coleman R.L., ed.). Wiley, New York, pp. 519–567.

Greenberg N. (1993). Central and endocrine aspects of tongue-flicking and exploratory behavior in *Anolis carolinensis*. *Brain Behav Evol* **41**, 210–218.

Griffith M. and Williams G. (1996). Roles of maternal vision and olfaction in suckling-mediated inhibition of luteinizing hormone secretion, expression of maternal selectivity, and lactational performance of beef cows. *Biol Reprod* **54**, 761–768.

Grigorjev C. and Munaro N. (1999). Time-dependent GABA-ergic activity in olfactory bulb and hypothalamus of proestrous rats. *Brain Res Bull* **48**, 569–572.

Guillamon A. and Segovia S. (1997). Sex differences in the vomeronasal system. *Brain Res Bull* **44**, 377–382.

Gulimova V.I. and Savelev S.V. (1996). Development of the *organon vomeronasalis* in human embryos of 19–40 mm crown-rump length. *Izvest Akad Nauk (Biol)* **2**, 185–192.

Guo J., Dudley C., Su T., Spink D., *et al.* (1999). Cytochrome P450 and steroid hydroxylase activity in mouse olfactory and vomeronasal mucosa. *Biochem Biophys Res Comm* **266**, 262–267.

Guo J., Zhou A. and Moss R. (1997). Urine and urine-derived compounds induce c-fos mRNA expression in accessory olfactory bulb. *Neuroreport* **8**, 1679–1683.

Guthman M.S., Harvey C. and Kling A. (1979). Somato-social development in the Rhesus monkey following olfactory bulbectomy. *Primates* **20**, 211–219.

Haas G. (1947). Jacobson's organ in the Chameleon. *J Morphol* **81**, 195–207.

Hagelin, L.-O. and Johnels, A. (1955). On the structure and function of the accessory olfactory organ in lampreys. *Acta Zool* **36**, 113–125.

Hagino-Yamagishi K., Matsuoka M., Wakabayashi Y., Mori Y. and Yazaki K. (2001). The mouse putative pheromone receptor was specifically activated by stimulation with male mouse urine. *J Biochem (Tokyo)* **129**, 509–512.

Haigh G., Cushing B.S. and Bronson F. (1988). A novel post-copulatory block to reproduction in White-footed mice. *Biol Reprod* **38**, 623–626.

Halem H., Cherry J. and Baum M. (1999). Vomeronasal neuroepithelium and forebrain fos responses to male pheromones in male and female mice. *J Neurobiol* **39**, 249–263.

Halpern M. (1992). Nasal chemical senses in reptiles: structure and function. In: *Biology of the Reptilia* (Gans C. and Crews D., eds.). Chicago University Press, pp. 423–524.

Halpern M., Halpern J., Erichsen E. and Borghjid S. (1997). The role of nasal chemical senses in garter snake response to airborne odor cues from prey. *J Comp Psychol* **111**, 251–260.

Halpern M., Jia C. and Shapiro L. (1998). Segregated pathways in the vomeronasal system. *Micros Res Tech* **41**, 519–529.

Halpern M. and Kubie J.L. (1984). The role of the ophidian vomeronasal system in species-typical behavior. *Trends Neurosci* **7**, 472–477.

Halpern M., Shapiro L. and Jia C. (1998). Heterogeneity in the accessory olfactory system. *Chem Senses* **23**, 477–481.

Halpern M., Shapiro L.S. and Jia C.P. (1995). Differential localisation of G-proteins in opossum vomeronasal system. *Brain Res* **677**, 157–161.

Hamlin H. (1929). Working mechanisms for the liquid and gaseous inake and output of Jacobson's organ. *Am J Physiol* **191**, 201–205.

Hansen A. and Finger T. (2000). Phyletic distribution of crypt-type olfactory receptor neurons in fishes. *Brain Behav Evol* **55**, 100–110.

Hansen A., Reiss J.O., Gentry C.L. and Burd G. (1998). Ultrastructure of the olfactory organ in the clawed frog, *Xenopus laevis* during larval development and metamorphosis. *J Comp Neurol* **398**, 273–288.

Hansen A. and Zeiske E. (1998). The peripheral olfactory organ of the zebrafish, *Danio rerio*: an ultrastructural study. *Chem Senses* **23**, 39–48.

Hansen A., Zeiske E. and Reutter K. (1994). Microvillous and ciliated receptor cells in the olfactory epithelium of the Australian Lungfish, *Neoceratodus forsteri*. In: *Advances in Biosciences* **93** (Apfelbach R., *et al.*, eds.). Elsevier, Oxford, pp. 43–51.

Hanson L., Sorensen P. and Cohen Y. (1998). Sex pheromones and amino acids evoke distinctly different spatial patterns of electrical activity in the goldfish olfactory bulb. *Ann NY Acad Sci* **855**, 521–524.

Hara T.J. (1994). Olfaction and gustation in fish: an overview. *Acta Physiol Scand* **152**, 207–217.

Harris M.A. and Murie J.O. (1984). Discrimination of estrous status by Columbian Ground Squirrels. *Anim Behav* **32**, 939–940.

Hart B.L. (1983). Flehmen behavior and vomeronasal organ function. In: *Chemical Signals in Vertebrates* **3** (Müller-Schwarze D. and Silverstein R.M., eds.). Plenum, New York, pp. 87–104.

Hart B.L. (1986). Medial preoptic-anterior hypothalamic lesions and sociosexual behavior of male goats. *Physiol Behav* **36**, 301–305.

Hart L.A. and Hart B.L. (1987). Species-specific patterns of urine investigation and Flehmen in Grant's Gazelle (*Gazella granti*), Thomson's Gazelle (*Gazella thomsoni*), Impala (*Aepyceros melampus*) and Eland (*Taurotragus oryx*). *J Comp Psychol* **101**, 299–304.

Hart B.L., Hart L. and Maina J.N. (1988). Alteration in vomeronasal system anatomy in Alcelaphine antelopes: correlation with alteration in chemosensory investigation. *Physiol Behav* **42**, 155–162.

Harteneck C., Plant T.D. and Schultz G. (2000). From worm to man: three subfamilies of TRP channels. *Trends Neurosci* **23**, 159–166.

Hatanaka T. (1991). Is the mouse vomeronasal organ a sex pheromone receptor? In: *Chemical Signals in Vertebrates* **6** (Doty R.L. and Müller-Schwarze D., eds). Plenum, New York, pp. 27–30.

Hatanaka T. and Matsuzaki O. (1993). Odor responses of the vomeronasal system in Reeve's turtle, *Geoclemys Reevesii. Brain Behav Evol* **41**, 183–186.

Hatt H., Gisselmann G., Wetzel C., *et al.* (1999). Cloning, functional expression and characterization of a human olfactory receptor. *Cell Mol Biol* **45**, 285–291.

Haug M. and Brain P. (1978). Attack directed by groups of castrated male mice towards lactating and non-lactating intruders: a urine-dependent phenomenon. *Physiol Behav* **21**, 549–552.

Hawes M. (1976). Odor as a possible isolating mechanism in sympatric species of shrews (*Sorex vagrans and S. obscurus*). *J Mammal* **57**, 404–406.

Hayashi S., Momiyama A., Ohishi H., Ogawameguro, R. *et al.* (1993). Role of metabotrophic glutamate receptor in synaptic modulation in accessory olfactory bulb. *Nature* **366**, 687–690.

Hedewig R. (1980). Vergleichende anatomische untersuchungen an den Jacobsonschen organen: 1, *Nycticebus coucang. Morphol Jb* **123**, 543–593.

Held W., Gallagher J.F., Hohman C.M., Kuhn N.J., *et al.* (1987). Identification and characterization of functional genes encoding the mouse major urinary proteins. *Mol Cell Biol* **10**, 3705–3712.

Hendrickx A.G. *et al.* (1971). *Embryology of the Baboon*. University of Chicago Press, Chicago, p. 325.

Hennessy M.B. (1980). Androgen related behavior in Squirrel monkey: an issue that is nothing to sneeze at. *Behav Neural Biol* **30**, 103–108.

Henzel W., Rodriguez H., Singer A.G., Stults J.T., *et al.* (1988). The primary structure of Aphrodisin. *J Biol Chem* **263**, 16682–16687.

Herrada G. and Dulac C. (1997). A novel family of putative pheromone receptors in mammals with a topographically organised and sexually dimorphic distribution. *Cell* **90**, 763–773.

Herrera E. and Macdonald D.W. (1994). Social significance of scent marking in Capybaras. *J Mammal* **75**, 410–415.

Herrick C. (1921). Connections of the vomeronasal organ, accessory olfactory bulb and amygdala in amphibia. *J Comp Neurol* **33**, 213–280.

Herzfeldt P. (1888). Uber das Jacobsonschen Organe des menschen und der Säugetiere. *Zool Jb* **3**, 551–574.

Heske E.J. and Nelson R.J. (1989). Pregnancy interruption in *Microtus ochrogaster* — laboratory artefact or field phenomenon? *Biol Reprod* **31**, 97–103.

Heth G., Nevo E., Ikan R., Weinstein V., *et al.* (1992). Differential olfactory perception of enantiomeric compounds by blind subterranean mole rats (*Spalax ehrenbergi*). *Experientia* **48**, 897–902.

Heth G. and Todrank J. (1995). Assessing chemosensory perception in subterranean mole rats: different responses to smelling versus touching odorous stimuli. *Anim Behav* **49**, 1009–1015.

Hildebrand J.G. and Shepherd G.M. (1997). Mechanism of olfactory discrimination: converging evidence for common principles across phyla. *Ann Rev Neurosci* **20**, 595–631.

Himstedt W. and Simon D. (1995). Sensory basis of foraging behaviour in Caecilians (Amphibia, Gymnophiona). *Herpetol J* **5**, 266–270.

Hofer H.O. (1978). The *ductus nasopalatinus* and the *ductus vomeronasalis*, and the occurrence of taste buds in the *papilla palatina* in *Nycticebus coucang* (Primates, Prosimiae). *VerhAnat Ges* **72**, 649–650.

Hofer H.O. (1980a). The external anatomy of the oro-nasal region of primates. *Z f Morphol u Anthropol* **71**, 233–249.

Hofer H.O. (1980b). Further observations on the occurrence of taste buds in the papilla palatina of primates. *Morphol Jb* **126**, 110–117.

Hofer H.O. (1982). Observation on the anatomy of the proboscis and of the ductus nasopalatinus and ductus vomeronasalis of *Solendon paradoxus* (Brandt, 1833). *Morphol Jb* **128**, 826–850.

Hofmann M. and Meyer D. (1995). The extrabulbar olfactory pathway — primary olfactory fibres by passing the olfactory bulb in bony fishes. *Brain Behav Evol* **46**, 378–388.

Hofmann T., Schaefer M., Schultz G. and Gudermann T. (2000a). Transient receptor potential channels as molecular substrates of receptor-mediated cation entry. *J Mol Med* **78**, 14–25.

Hofmann T., Schaefer M., Schultz G. and Gudermann T. (2000b). Cloning, expression and subcellular localisation, of two novel splice variants of mouse transient receptor potential channel-2. *Biochem J* **351**, 115–122.

Hollnagel-Jensen O.C. and Andreasen E. (eds.) (1948). Preface and notes: Ouvrages sur l'Organe voméro-nasal, etc.; Einar Munksgaard, Copenhague, xxxii & ill., pp. 175. Einar Munksgaard, Kobenhaven, pp. 163.

Holmes D. (1992). Odors as cues for orientation to mothers by weanling Virginia Opossums. *J Chem Ecol* **18**, 2251–2259.

Holtzman D.A. and Halpern M. (1990). Embryonic and neonatal development of the vomeronasal and olfactory systems in garter snakes. *J Morphol* **203**, 123–140.

Holy T., Dulac C. and Meister M. (2000). Responses of vomeronasal neurons to natural stimuli. *Science* **289**, 1569–1572.

Hoon M., Adler E., Lindemeier J., Battey J.F., *et al.* (1999). Putative mammalian taste receptors: a class of taste-specific GPCRs with distinct topographic selectivity. *Cell* **96**, 541–551.

Hoppe P. (1975). Genetic and endocrine studies of the pregnancy-blocking pheromones of mice. *J Reprod Fertil* **45**, 109–115.

Horowitz L., Montmayeur J.P., Echelard Y. and Buck L. (1999). A genetic approach to trace neural circuits. *Proc Natl Acad Sci* **96**, 3194–3199.

Howes G. (1891). On the probable existence of a Jacobson's Organ among the *Crocodilia* etc. *Proc Zool Soc Lond*, 148–158.

Huber G. and Guild R. (1913). Observations on the peripheral distribution of the *Nervus terminalis* in Mammalia. *Anat Rec* **7**, 253–272.

Hudson R. and Distel H. (1986). Pheromonal release of suckling in rabbits does not depend on the vomeronasal organ. *Physiol Behav* **37**, 123–129.

Hummel T., Kühnau D., Knecht M., Abolmaali N. and Hüttenbrink K.B. (1999). The anatomy of the vomeronasal organ: characterisation by means of nasal endoscopy and magnetic resonance imaging. *Chem Senses* **24**, abs. 365.

Humphrey T. (1940). The development of the olfactory and accessory olfactory formation in human embryos and fetuses. *J Comp Neurol* **73**, 431–468.

Hunter A.J., Fleming D. and Dixson A. (1984). The structure of the vomeronasal organ and nasopalatine ducts in *Aotus trivirgatus* and some other primate species. *J Anat* **138**, 217–225.

Hurst J.L. (1987). Function of urine-marking in a free-living population of house mice, *Mus domesticus. Anim Behav* **35**, 1433–1442.

Hurst J.L. (1989). The complex network of olfactory communication in populations of wild house mice *Mus domesticus*: urine marking and investigation within family groups. *Anim Behav* **37**, 705–725.

Hurst J.L. and Nevison C. (1994). Do female house mice, *Mus domesticus*, regulate their exposure to reproductive priming pheromones? *Anim Behav* **48**, 945–959.

Hurst J.L., Robertson D., Tolladay U. and Beynon R. (1998). Proteins in urine scent marks of male house mice extend the longevity of olfactory signals. *Anim Behav* **55**, 1289–1297.

Ichikawa M. (1988). Plasticity of intra-amygdaloid connections following the denervation of fibres from accessory olfactory bulb to medial amygdaloid nucleus in adult rat. *Brain Res* **451**, 248–254.

Ichikawa M. (1999). Axonal growth of newly formed vomeronasal receptor neurons after nerve transection. *Anat Embryol* **200**, 413–417.

Ichikawa M., Matsuoka M. and Mori K. (1995). Plastic effects of soiled bedding on the structure of synapses in rat accessory olfactory bulb. *Synapse* **21**, 104–109.

Ichikawa M. and Oka Y. (1988). Increase of LHRH immunoreactive fibres in medial amygdaloid nucleus after removal of accessory olfactory bulb. *Zool Sci (Tokyo)* **5**, 1300–1312.

Ichikawa M., Takami S., Osada T. and Graziadei P.P.C. (1994). Differential development of binding-sites of 2 lectins in the vomeronasal axons of the rat accessory olfactory bulb. *Dev Brain Res* **78**, 1–9.

Iida A. and Kashiwayanagi M. (1999). Responses of *Xenopus laevis* water nose to water-soluble and volatile odorants. *J Gen Physiol* **114**, 85–92.

Iida A. and Kashiwayanagi M. (2000). Responses to putative second messengers and odorants in water nose olfactory neurons of *Xenopus laevis*. *Chem Senses* **25**, 55–59.

Inamura K. and Kashiwayanagi M. (2000a). Inhibition of fos-immunoreactivity in response to urinary pheromones by beta-adrenergic and serotonergic antagonists in the rat accessory olfactory bulb. *Biol Pharm Bull* **23**, 1108–1110.

Inamura K. and Kashiwayanagi M. (2000b). Inward current responses to urinary substances in rat vomeronasal sensory neurons. *Eur J Neurosci* **12**, 3529–3536.

Imamura K., Kashiwayanagi M. and Kurihara K. (1997). Blockage of urinary responses by inhibitors for IP-3 mediated pathway in rat vomeronasal sensory neurons. *Neurosci Lett* **233**, 129–132.

Inamura K., Kashiwayanagi M. and Kurihara K. (1999). Regionalization of Fos immunostaining in rat accessory olfactory bulb when the vomeronasal organ was exposed to urine. *Eur J Neurosci* **11**, 2254–2256.

Inamura K., Matsumoto Y., Kashiwayanagi M. and Kurihara K. (1999). Laminar distribution of pheromone-receptive neurons in rat vomeronasal epithelium. *J Physiol* **517**, 731–739.

Inouchi J., Kubie J. and Halpern M. (1989). Accessory olfactory bulb neurones respond to liquid odourants delivered to vomeronasal organ. *Chem Senses* **14**, 712.

Issel-Tarver L. and Rine J. (1996). Organization and expression of canine olfactory receptor genes. *Proc Natl Acad Sci* **93**, 10897–10902.

Iwahori N., Nakamura K. and Mameya C. (1989). A Golgi study on the accessory olfactory bulb in the snake *Elaphe quadrivirgata*. *Neurosci Res* **7**, 55–70.

Iwata T., Umezawa K., Toyoda F., Takahashi N., *et al.* (1999). Molecular cloning of newt sex pheromone precursor cDNAs: evidence for the existence of species-specific forms of pheromones. *FEBS Lett* **457**, 400–404.

Izard M.K. (1990). Social influences on the reproductive success and reproductive endocrinology of prosimian primates. In: *Socioendocrinology of Primate Reproduction* (Bercovitch F. and Ziegler T., eds.). Wiley-Liss, New York, pp. 159–186.

Izard M.K. and Vandenbergh J.G. (1982a). Priming pheromones from oestrous cows increase synchronisation of estrous in dairy heifers after PGF-2α injection. *J Reprod Fertil* **66**, 189–196.

Izard M.K. and Vandenbergh J.G. (1982b). Effects of bull urine on puberty and calving date in crossbred heifers. *J Anim Sci* **55**, 1160–1168.

Izquierdo M.A., Collado P., Segovia S., Guillamon A. and del Cerro M. (1992). Maternal behavior induced in male rats by bilateral lesions of the bed nucleus of the accessory olfactory tract. *Physiol Behav* **52**, 707–712.

Jackson L.M. and Harder J.D. (1996). Vomeronasal organ removal blocks pheromonal induction of estrus in Gray Short-tailed Opossums (*Monodelphis domestica*). *Biol Reprod* **54**, 506–512.

Jacobs V.L., Sis R.F., Chenoweth P.J., Klemm W.R. and Sherry C.J. (1981). Structures of the bovine vomeronasal complex and its relationship to the palate. *Acta Anat* **110**, 48–58.

Jacobs V.L., Sis R.F., Chenoweth P.J., Klemm W.R., *et al.* (1980). Tongue manipulation of the palate assists estrous detection in the bovine. *Theriogenol* **13**, 353–356.

Jacobson Louis, Description anatomique d'un organe observé dans les Mammifères (*Mém.*, remis a' l'Institut de France: le 12 Janvier, 1812).

Jaeger R. and Gergits W. (1979). Inter- and intra-specific communication in salamanders through chemical signals on the substrate. *Anim Behav* **27**, 150–156.

Jahnke V. and Merker H. (1998). Elektronenmikroskopische Untersuchungen des menschlichen vomeronasalen Organs. *HNO* **46**, 502–506.

Jansen H., Iwamoto G. and Jackson G. (1998). Central connections of the ovine olfactory bulb formation identified using wheat germ agglutinin-conjugated horseradish peroxidase. *Brain Res Bull* **45**, 27–39.

Janus C. (1988). The development of responses to naturally occurring odors in Spiny Mice, *Acomys cahirinus*. *Anim Behav* **36**, 1400–1406.

Janus C. and Holman S. (1989). Development of sex differences in the response of Spiny Mouse pups to adult male odors. *Physiol Behav* **46**, 895–900.

Jarvis J.U.M. (1991). Reproduction of naked mole rats. In: *The Biology of the Naked Mole-Rat* (Sherman P., Jarvis M. and Alexander R., eds.). Princeton University Press, pp. 384–325.

Jastrow H. and Oelschlager H. (1998). Anatomy and development of the terminal Nv. and Jacobson's Organ. *Ann Anat* **181**(Suppl.), 75.

Jemiolo B., Bozena F.A., Xie T., Wiesler D. and Novotny M. (1989). Puberty-affecting synthetic analogs of urinary chemosignals in the house mouse (*Mus domesticus*). *Physiol Behav* **46**, 293–298.

Jemiolo B., Gubernick D.J., Yoder M.C. and Novotny M. (1994). Chemical characterization of urinary volatile compounds of *Peromyscus californicus*, a monogamous biparental rodent. *J Chem Ecol* **20**, 2489–2500.

Jemiolo B. and Novotny M. (1994). Inhibition of sexual maturation in juvenile female and male mice by a chemosignal of female origin. *Physiol Behav* **55**, 519–522.

Jia C., Chen W.R. and Shepherd G. (1999). Synaptic organization and neurotransmitters in the rat accessory olfactory bulb. *J Neurophysiol* **81**, 345–355.

Jia C., Goldmann J. and Halpern M. (1997). Development of vomeronasal receptor neurons subclasses and establishment of topographic projection to accessory olfactory bulb. *Dev Brain Res* **102**, 209–216.

Jia C. and Halpern M. (1996). Subclasses of vomeronasal receptor neurons: differential expression of G-proteins (Gi-alpha2 & Go-alpha) and segregated projections to the accessory olfactory-bulb. *Brain Res* **719**, 117–128.

Jia C. and Halpern M. (1998). Neurogensis and migration of receptor neurons in the vomeronasal sensory epithelium of the Opossum *Monodelphis domestica. J Comp Neurol* **400**, 287–297.

Jia C.P. and Halpern M. (1997). Segregated populations of mitral/tufted cells in the accessory olfactory bulb. *Neuroreport* **8**, 1887–1890.

Jiang X.C., Inouchi J., Wang D. and Halpern M. (1990). Purification and characterisation of a chemoattractant from earthworm electric-shock-induced secretion and vomeronasal system in Garter snakes. *J Biol Chem* **265**, 8736–8744.

Johns M.A. (1978). Urine-induced reflex ovulation in anovulatory rats may be a vomeronasal effect. *Nature* **272**, 446–448.

Johnson A.R., Josephson R. and Hawke M. (1985). Clinical and histological evidence for the presence of the vomeronasal (Jacobson's) organ in adult humans. *J Otolaryngol* **14**, 71–79.

Johnson E.W., Eller P.M., Jafek B.W. and Norman A.W. (1992). Calbindin-like immunoreactivity in two peripheral chemosensory tissues of the rat: taste buds and the vomeronasal organ. *Brain Res* **572**, 319–324.

Johnson E.W., Eller P. and Jafek B.W. (1993). An immunoelectron microscopic comparison of olfactory marker protein localization in the supranuclear regions of the rat olfactory epithelium and vomeronasal organ neuroepithelium. *Acta Oto-Laryngol* **113**, 766–771.

Johnson E.W., Eller P.M. and Jafek B.W. (1994a). Calbindin-like immunoreactivity in epithelial cells of newborn and adult human vomeronasal organ. *Brain Res* **638**, 329–333.

Johnson E.W., Eller P. and Jafek B.W. (1994b). Protein gene-product-9.5 in the developing and mature rat vomeronasal organ. *Dev Brain Res* **78**, 259–264.

Johnson E.W., Eller P. and Jafek B. (1995). Distribution of OMP, PGP 9.5- and CaBP-like immunoreactive chemoreceptor neurons in the developing human olfactory epithelium. *Anat Embryol* **191**, 311–317.

Johnston R.E. (1985). Olfactory and vomeronasal mechanisms of communication. In: *Taste, Olfaction and the Central Nervous System* (Pfaff D.W., eds.). Rockefeller University Press, N.Y.

Johnston R.E. (1992a). Olfactory and vomeronasal mechanisms of social communication in hamsters. In: *Chemical Signals in Vertebrates* **6** (Doty R.L. and Müller-Schwarze D., eds). Plenum, New York, pp. 49–54.

Johnston R.E. (1992b). Vomeronasal and or olfactory mediation of ultrasonic calling and scent marking by female golden hamsters. *Physiol Behav* **51**, 437–448.

Johnston R.E. (1998). Pheromones, the vomeronasal system and communication — from hormonal responses to individual recognition. *Ann NY Acad Sci* **855**, 333–348.

Johnston R.E. (2000). Chemical communication and pheromones: the types of chemical signals and the role of the vomeronasal system. In: *The Neurobiology of Taste and Smell*, 2nd ed. (Finger T.E., Silver W.L. and Restrepo D., eds.). Plenum, New York, pp. 101–128.

Johnston R.E. and Rasmussen K. (1984). Individual recognition of female hamsters by males: role of chemical cues and of the olfactory and vomeronasal systems. *Physiol Behav* **33**, 95–104.

Johnston R.E. and Peng M. (2000). The vomeronasal organ is involved in discrimination of individual odors by males but not by females in golden hamsters. *Physiol Behav* **70**, 537–549.

Jones F., Pfeiffer C. and Asashima M. (1994). Ultrastructure of the olfactory organ of the newt, *Cynops pyrrhogaster*. *Ann Anat* **176**, 269–275.

Josefsson L. (1999). Evidence for kinship between diverse G-protein coupled receptors. *Gene* **239**, 333–340.

Kaba H. and Kawasaki Y. (1996). A one-dimensional current source-density analysis is applicable to the mouse accessory olfactory bulb. *J Vet Med Sci* **58**, 485–488.

Kaba H., Li C., Keverne E.B., Saito H. and Seto K. (1992). Physiology and pharmacology of the accessory olfactory system. In: *Chemical Signals in Vertebrates* **6** (Doty R.L., *et al.*, eds.). Plenum, New York, pp. 49–54.

Kaba H. and Nakanishi S. (1995). Synaptic mechanisms of olfactory recognition memory. *Int Rev Neurosci* **6**, 125–141.

Kaba H., Rosser A. and Keverne E.B. (1988). Hormonal enhancement of neurogenesis and its relationship to the duration of olfactory memory. *Neuroscience* **24**, 93–98.

Kahmann H. (1932). Sinnesphysiologischen Studien an Reptilien, I. Expt. Untersuchungen über das Jacobsonsches Organe der Eidechsen u Schlangen. *Zool Jb* **51**, 173–238.

Kaneko N., Debski E.A., Wilson M.C. and Whitten W.K. (1980). Puberty acceleration in mice: II, Evidence that the vomeronasal organ is a receptor for the primer pheromone in male mouse urine. *Biol Reprod* **22**, 873–878.

Kapusta J., Marchlewska-Koj A., Olejnicza K. and Kruczek M. (1996). Removal of the olfactory system modifies male Bank Vole behavior in the presence of females. *Behav Proc* **37**, 39–45.

Kauer J.S. (1991). Contributions of topography and parallel processing to odor coding in the vertebrate olfactory pathway. *Trends Neurosci* **14**, 79–85.

Keil W. and von Stralendorff F. (1990). A behavioral bioassay for analysis of rabbit nipple-search pheromone. *Physiol Behav* **47**, 525–530.

Kelche C. and Aron C. (1984). Olfactory cues and accessory olfactory bulb lesion: effect on sexual behavior in the cyclic female rat. *Physiol Behav* **33**, 45–48.

Kelliher K., Chang Y., Wersinger S. and Baum M. (1998). Sex difference and testosterone modulation of pheromone-induced neuronal *fos* in the Ferret's main olfactory bulb and hypothalamus. *Biol Reprod* **59**, 1454–1463.

Kelly D. (1996). When is a butterfly like an elephant? *Chem Biol* **3**, 595–602.

Kemble E.D. and Nagel J.A. (1975). Decreased sniffing behavior in rats following septal lesions. *Psychonom Sci* **5**, 309–310.

Kendrick K.M., DaCosta A., Broad K.D., Ohkura S., *et al.* (1997). Neural control of maternal behaviour and olfactory recognition of offspring. *Brain Res Bull* **44**, 383–395.

Kendrick K.M., Guevara-Guzman R., Zorrilla J., Hinton M., *et al.* (1997). Formation of olfactory memories mediated by nitric oxide. *Nature* **388**, 670–674.

Keverne E.B. (1979). The dual olfactory projections and their significance for behavior. In: *Chemical Ecology: Odour Communication in Animals* (Ritter F.J., ed.). Elsevier, Amsterdam, pp. 75–83.

Keverne E.B. (1983). Pheromonal influences on the endocrine regulation of reproduction. *Trends Neurosci* **6**, 381–384.

Keverne E.B. (1999). Vomeronasal organ. *Science* **286**, 716–720.

Keverne E.B. and de la Riva C. (1982). Pheromones in mice: reciprocal interaction between the nose and brain. *Nature* **296**, 148–150.

Keverne E.B. and Kaba H. (1990). A neural mechanism for olfactory learning. In: *Chemical Signals in Vertebrates* **5** (MacDonald D., Müller-Schwarze D. and Natynczuk S.E., eds.). Oxford University Press, Oxford, pp. 87–99.

Kevetter G.A. and Winans S. (1981). Connections of the cortico-medial amygdala in the golden hamster *Mesocricetus auratus*, 1. Efferents of the 'vomeronasal amygdala'. *J Comp Neurol* **197**, 81–98.

Key S. and Wray S. (2000). Two olfactory placode derived galanin subpopulations: luteinizing hormone-releasing hormone neurones and vomeronasal cells. *J Neuroendocrinol* **12**, 535–545.

Khew-Goodall Y., Grillo M., Getchell M., Danho W., *et al.* (1991). Vomeromodulin, a putative pheromone transporter: cloning, characterization, and cellular-localization of a novel glycoprotein of lateral nasal gland. *FASEB J* **5**, 2976–2982.

Kikuyama S., Toyoda F., Ohmiya Y., Matsuda K., *et al.* (1995). Sodefrin: a female-attracting peptide hormone in newt cloacal glands. *Science* **267**, 1643–1645.

Kikuyama S., Toyoda F., Yamamoto K., Tanaka S., *et al.* (1997). Female-attracting pheromone in newt cloacal glands. *Brain Res Bull* **44**, 415–422.

Kim H. and Greer C. (2000). The emergence of compartmental organization in olfactory bulb glomeruli during postnatal development. *J Comp Neurol* **422**, 297–311.

Kim K., Patel L., Tobet S.A., King J.C., *et al.* (1999). Gonadotropin-releasing hormone immunoreactivity in the adult and fetal human olfactory system. *Brain Res* **826**, 220–229.

Kirkpatrick B., Carter C., Newman S. and Insel T. (1994). Axon-sparing lesions of the medial nucleus of the amygdala decrease affiliative behaviors in the prairie vole (*Microtus ochrogaster*) — behavioral and anatomical specificity. *Behav Neurosci* **108**, 501–513.

Kishimoto J., Keverne E., Hardwick J. and Emson P. (1993). Localization of nitric oxide synthase in the mouse olfactory and vomeronasal system — a histochemical, immunological and *in-situ* hybridization study. *Europ J Neurosci* **5**, 1684–1694.

Kjaer I. and Hansen B. (1996). The human vomeronasal organ: prenatal developmental stages and distribution of luteinizing-hormone releasing hormone. *Eur J Oral Sci* **104**, 1684–1694.

Klauer G. (1984). Macroscopic and microscopic anatomy of the external nose in *Tarsius bancanus*. In: *Biology of Tarsiers* (Niemitz C., ed). G. Fischer, Stuttgart, pp. 291–302.

Klemm W.R., Sherry C.J., Sis R.F. and Morris D.L. (1984). Electrographic recording from bovine vomeronasal capsule under spontaneous and stimulated conditions. *Brain Res Bull* **12**, 275–282.

Klemm W.R., Hawkins G. and de Los Santos E. (1987). Identification of compounds in bovine cervico-vaginal mucous extracts that evoke male sexual behavior. *Chem Senses* **12**, 77–87.

Kligman A. and Shehadeh N. (1964). Pubic apocrine glands and odor. *Arch Dermatol* **89**, 461–464

Knappe H. (1964). Zur Funktion des Jacobsonschen Organs (*Organon vomeronasale jacobsonii*). *Zool Garten (NF)* **28**, 188–194.

Kollack-Walker S. and Newman S.W. (1997). Mating-induced expression of c-fos in the male Syrian hamster brain: role of experience, pheromones, and ejaculations. *Neurobiol J* **32**, 481–501.

Kolmer W. (1927). Geruchsorgane. In: *Handbuch der Mikroskop. Anatomie des Menschen* (von Molendorff, ed.). Vol. 3/1, pp. 192–249.

Kolnberger I. (1971). Comparative studies of the olfactory epithelium especially the Vomeronasal (Jacobson's) Organ in Amphibia, Reptiles and Mammals. *Z Zellf Mikrosk Anat* **122**, 53–67.

Kolunie J. and Stern J. (1995). Maternal aggression in rats — effects of olfactory bulbectomy, $ZnSO_4$-induced anosmia, and vomeronasal organ removal. *Horm Behav* **29**, 492–518.

Kosaka T. and Hama K. (1982). Synaptic organisation of the teleost olfactory bulb. *J Physiol (Paris)* **78**, 707–719.

Kosmider J. (1990). Olfactory adaptation: a factor limiting warning role of odor. *Med Pr* **41**, 142–147.

Kotrschal K. (2000). Taste(s) and olfaction(s) in fish: a review of specialized subsystems and central integration. *Pflugers Arch* **439**(3, Suppl.), 78–80.

Kratzing J.E. (1971a). The fine structure of the sensory epithelium of the vomeronasal organ in suckling rats. *Aust J Biol Sci* **24**, 787–796.

Kratzing J.E. (1971b). The structure of the vomeronasal organ in the sheep. *J Anat* **108**, 247–260.

Kratzing J.E. (1975). The fine structure of the olfactory and vomeronasal organs of a lizard (*Tiliqua scincoides*). *Cell Tissue Res* **156**, 239–252.

Kratzing J.E. (1978). The olfactory apparatus of the Bandicoot (*Isodon macrourus*). *J Anat* **125**, 601–613.

Kratzing J.E. (1980). Unusual features of the vomeronasal organ of young pigs. *J Anat* **130**, 213 abs.

Kratzing J.E. (1982). The anatomy of the rostral nasal cavity and vomeronasal organ in *Tarsipes rostratus* (Marsupialia: Tarsipedidae). *Aust Mammal* **5**, 211–219.

Kratzing J.E. (1984). The anatomy and histology of the nasal cavity of the Koala (*Phascolarctus cinereus*). *J Anat* **138**, 55–65.

Kratzing J.E. (1986). Morphological maturation of the olfactory epithelia of Australian marsupials. In: *Ontogeny of Olfaction* (Breipohl W., ed.). Springer, Berlin, pp. 57–70.

Kratzing J.E. (1987). The presence of rostral palatal taste buds in mammals. *Aust Mammal* **10**, 29–32.

Kratzing J.E. and Woodall P. (1988). The rostral nasal anatomy of two Elephant Shrews. *J Anat* **157**, 135–143.

Kreutzer E.W. and Jafek B.W. (1980). The vomeronasal organ of Jacobson in the human embryo and fetus. *Otol HeadNeck Surg* **88**, 119–123.

Krieger J., Schmitt A.L., Gudermann T., Schultz G., *et al.* (1999). Selective activation of G-protein subtypes in the vomeronasal organ upon stimulation with urine-derived compounds. *J Biol Chem* **274**, 4655–4662.

Krishna N.S., Getchell T.V. and Getchell M.L. (1992). Differential distribution of gamma-glutamyl cycle molecules in the vomeronasal organ of rats. *Neurorep* **3**, 551–554.

Krishna R.N., Getchell T.V. and Getchell M.L. (1994). Differential expression of alpha-class, mu-class, and pi-class of glutathione S-transferases in chemosensory mucosae of rats during development. *J Neurosci Res* **39**, 243–259.

Krishna N., Getchell T.V., Margolis F. and Getchell M.L. (1992). Amphibian olfactory receptor neurons express olfactory marker protein. *Brain Res* **593**, 295–298.

Krishna N., Getchell M., Margolis F. and Getchell T. (1995). Differential expression of vomeromodulin and odorant-binding protein, putative pheromone and odorant transporters, in the developing rat nasal chemosensory mucosae. *J Neurosci Res* **40**, 54–71.

Kroner R., Breer H., Singer A. and O'Connell R. (1996). Pheromone-induced second messenger signaling in the hamster vomeronasal organ. *Neurorep* **7**, 2989–2992.

Kruczek M. and Marchlewska-Koj A. (1986). Puberty delay of Bank Vole females in a high-density population. *Biol Reprod* **35**, 537–541.

Kruhoffer M., Bub A., Cieslak A., Adermann K., *et al.* (1997). Gene expression of aphrodisin in female hamster genital tract segments. *Cell Tissue Res* **287**, 153–160.

Krzymowski T., Grzegorzewski W., Stefanczyk-Krzymowska S., Skipor J., *et al.* (1999). Humoral pathway for transfer of the boar pheromone androstenol, from the nasal mucosa to the brain and hypophysis of gilts. *Theriogenol* **52**, 1225–1240.

Kruzhalov N.B. (1980). Neuronal reactions of the accessory olfactory bulb in the frog, *Rana temporaria*, to chemical stimulation of the vomeronasal organ. *Zh Evol Biokhim Fizio* **16**, 587–592.

Kubie J.L. and Halpern M. (1979). The chemical senses involved in Garter snake prey trailing. *J Comp Physiol Psychol* **93**, 648–667.

Kubie J.L., Vagvolgyi A. and Halpern M. (1978). Roles of the vomeronasal and olfactory systems in courtship behavior of male Garter snakes. *J Comp Physiol Psychol* **92**, 627–641.

Kuhn N., Woodworth-Gutai M., Gross K. and Held W. (1984). Subfamilies of MUP multi-gene family: sequence analysis of cDNA clones and diffferential regulation in liver. *Nucleic Acid Res* **12**, 6073–6090.

Kulkarni A., Getchell T. and Getchell M. (1994). Neuronal nitric oxide synthase is localized in extrinsic nerves regulating perireceptor processes in the chemosensory nasal mucosae of rats and humans. *J Comp Neurol* **345**, 125–138.

Kulkarni-Narla A., Getchell T.V. and Getchell M.L. (1997). Differential expression of manganese and copper-zinc superoxide dismutases in the olfactory and vomeronasal receptor neurons of rats during ontogeny. *J Comp Neurol* **381**, 31–40.

Kumar A., Dudley C. and Moss R. (1999). Functional dichotomy within the vomeronasal system: distinct zones of neuronal activity in the accessory olfactory bulb correlate with sex-specific behaviors. *J Neurosci* **19**, 1–6.

Kuser P., Krauchenco S., Fangel A. and Polikarpov T. (1999). Crystallization and preliminary diffraction studies of a recombinant major urinary protein. *Acta Crystallogr (D)* **55**, 1340–1341.

Labov J.B. and Wysocki C.J. (1989). Vomeronasal organ and social factors affect urine marking by male mice. *Physiol Behav* **45**, 443–447.

Labov J.B. Katz Y., Wysocki C.J., Beauchamp G., *et al.* (1988). Levels of immunoreactive β-endorphin in rostral and caudal sections of olfactory bulbs from male guinea pigs exposed to odors of conspecific females. *Ann NY Acad Sci* **510**, 432–435.

Lacalli T. and Hou S. (1999). A re-examination of the epithelial sensory cells of Amphioxus (*Branchiostoma*). *Acta Zool* **80**, 125–134.

Ladewig J., Price E.O. and Hart B.L. (1980). Flehmen in male goats: role in sexual behavior. *Behav Neural Biol* **30**, 312–322.

Lai S., Vasilieva N.Y. and Johnston R.E. (1996). Odors providing sexual information in Djungarian hamsters: evidence for an across-odor code. *Horm Behav* **30**, 26–36.

Lane R., *et al.* (2002). Sequence analysis of mouse vomeronasal receptor gene clusters reveals common promoter motifs and a history of recent expansion. *Proc Natl Acad Sci* **99**, 291–296.

Lanthier A. and Patwardhan V.V. (1987). Effect of heterosexual olfactory and visual stimulation on 5α-en-3β hydroxysteroids and progesterone in the male rat brain. *J Steroid Biochem* **28**, 697–701.

Lanuza E. and Halpern M. (1997). Afferent and efferent connections of the nucleus sphericus in the snake *Thamnophis sirtalis*: convergence of olfactory and vomeronasal information in the lateral cortex and the amygdala. *J Comp Neurol* **385**, 627–640.

Lanuza E. and Halpern M. (1998). Efferents and centrifugal afferents of the main and accessory olfactory bulbs in the snake *Thamnophis sirtalis*. *Brain Behav Evol* **51**, 1–22.

Larriva-Sahd J.A., Matsumoto A. and Sumoto A. (1994). The vomeronasal system and its connections with sexually dimorphic neural structures. *Zool Sci* **11**, 495–506.

Larsell O. (1950). Studies on the *Nervus terminalis*: mammals. *J Comp Neurol* **130**, 3–68.

Laska M. and Hudson R. (1995). Ability of female Squirrel Monkeys to discriminate between conspecific urine odours. *Ethology* **99**, 39–52.

Lau Y. and Cherry J. (2000). Distribution of PDE4A and G(o)-alpha immunoreactivity in the accessory olfactory system of the mouse. *Neurorep* **11**, 27–32.

Laukaitis C.M., Critser E.S. and Karin R.C. (1997). Salivary androgen-binding protein (ABP) mediates sexual isolation in *Mus musculus. Evolution* **51**, 2000–2005.

Lawton A.D. and Whitsett J.M. (1979). Inhibition of sexual maturation by a urinary pheromone in male prairie deer mice. *Horm Behav* **13**, 128–138.

Lee C. and Ingersoll D. (1979). Salivary cues in the mouse: a preliminary study. *Horm Behav* **12**, 20–29.

Lee K.-H., Wells R.G. and Reed R. (1987). Isolation of an olfactory complementary DNA: similarity to retinol-binding protein suggests a role in olfaction. *Science* **253**, 1053–1066.

Lefcort H. (1996). Adaptive, chemically mediated fright response in tadpoles of the Southern Leopard frog, *Rana utricularia. Copeia* **2**, 455–459.

Legouis R., Cohen-Salmon M., Delcastillo I. and Petit C. (1994). Isolation and characterization of the gene responsible for the X-chromosome linked Kallmann syndrome. *Biomed Pharm* **48**, 241–246.

Lehman M.N. and Winans S.S. (1982). Vomeronasal and olfactory pathways to the amygdala controlling male hamster sexual behavior — autoradiographic and behavioral analyses. *Brain Res* **240**, 27–41.

Lehman M.N., Newman S. and Silverman A.-J. (1987). LHRH in the vomeronasal system and the Terminal nerve of the Hamster. *Ann NY Acad Sci* **519**, 229–240.

Lehman-McKeeman L., Caudill D., Rodriguez P. and Eddy C. (1998). 2-sec-butyl-4,5-dihydrothiazole is a ligand for mouse urinary protein and rat alpha 2μ-globulin: physiological and toxicological relevance. *Toxicol Appl Pharmacol* **149**, 32–40.

Leinders-Zufall T., Lane A.P., Puche A.C., Ma W., *et al.* (2000). Ultrasensitive pheromone detection by mammalian vomeronasal neurons. *Nature* **405**, 792–796.

Lepri J.J. (1994). The vomeronasal organ coordinates reproduction in some, but not all, Prairie Voles. In: *Advances in Biosciences* **93** (Apfelbach R., *et al.*, eds.). Pergamon/ Oxford, pp. 273–280.

Lepri J.L. and Wysocki C.J. (1987). Removal of the vomeronasal organ disrupts the activation of reproduction in female voles. *Physiol Behav* **40**, 349–356.

Lepri J.L. and Wysocki C.J. (1991). Consequences of removing the vomeronasal organ. *J Ster Biochem Molec Biol* **39**(4b), 661–670.

Lepri J.J., Wysocki C.J. and Vandenbergh, J.G. (1985). Mouse vomeronasal organ: effects on chemosignal production and maternal behavior. *Physiol Behav* **35**, 809–814.

Levy F., Locatelli A., Piketty V., Tillet Y. and Poindron P. (1995). Involvement of the main but not the accessory olfactory system in maternal behavior of primiparous and multiparous ewes. *Physiol Behav* **57**, 97–104.

Levy F. and Poindron P. (1987). The importance of amniotic fluid for the establishment of maternal behavior in experienced and inexperienced ewes. *Anim Behav* **35**, 1188–1192.

Leyden J., McGinley K., Hoelzle K., Labows J., *et al.* (1981). The microbiology of the human axillae in relation to axillary odors. *J Invest Dermatol* **77**, 413–416.

Li C., Kaba H., Saito H. and Seto K. (1990). Neural mechanisms underlying the action of primer pheromones in mice. *Neuroscience* **36**, 773–778.

Li C., Kaba H. and Seto K. (1994). Effective induction of pregnancy block by electrical-stimulation of the mouse accessory olfactory-bulb coincident with prolactin surges. *Neurosci Lett* **176**, 5–8.

Li C.S., Kaba H., Saito H. and Seto K. (1992a). Cholecystokinin: critical role in mediating olfactory influences on reproduction. *Neuroscience* **48**, 707–713.

Li C.S., Kaba H., Saito H. and Seto K. (1992b). Estrogen infusions into the amygdala potentiate excitatory transmission from the accessory olfactory bulb to tuberoinfundibular arcuate neurons in the mouse. *Neurosci Lett* **143**, 48–50.

Licht G. and Meredith M. (1987). Convergence of main and accessory olfactory pathways onto single neurons in the hamster amygdala. *Exp Brain Res* **69**, 7–18.

Liman E., Corey D. and Dulac C. (1999). TRP-2: a candidate transduction channel for mammalian pheromone sensory signaling. *Proc Natl Acad Sci* **96**, 5791–5796.

Lin D.M. and Ngai J. (1999). Development of the vertebrate main olfactory system. *Curr Opin Neurobiol* **9**, 74–78.

Lindsay F.E.F. and Burton F. (1984). Observational study of urine-testing in the horse and donkey stallion. *Equine Vet J* **15**, 330–336.

Liu J., Chen P., Wang D. and Halpern M. (1999). Signal transduction in the vomeronasal organ of garter snakes: ligand-receptor binding mediated protein phosphorylation. *Biochem Biophys Acta* **1450**, 320–330.

Liu W.M., Wang D., Liu J., Chen P. and Halpern M. (1998). Chemosignal transduction in the vomeronasal organ of garter snakes: cloning of a gene encoding adenylate cyclase from the vomeronasal organ of garter snakes. *Arch Biochem Biophys* **358**, 204–210.

Llahi S. and Garcia-Verdugo J.M. (1989). Neuronal organization of the accessory olfactory bulb of the lizard *Podarcis hispanica*: a Golgi study. *J Morphol* **202**, 13–28.

Lobel D., Marchese S., Krieger J., Pelosi P., *et al.* (1998). Subtypes of odorant-binding proteins: heterologous expression and ligand binding. *Europ J Biochem* **254**, 318–324.

Loebel D., Scaloni A., Paolini S., Fini C., *et al.* (2000). Cloning, post-translational modifications, heterologous expression and ligand-binding of boar salivary lipocalin. *Biochem J* **350**, 369–379.

Lohman A. and Smeets W. (1993). Overview of the main and accessory olfactory bulb projections in reptiles. *Brain Behav Evol* **41**, 147–155.

Lomas D.E. and Keverne E.B. (1982). Role of vomeronasal organ and prolactin in the acceleration of puberty in female mice. *J Reprod Fert* **66**, 101–107.

Lucke C., Franzoni L., Abbate F., Lohr F., *et al.* (1999). Solution structure of a recombinant mouse major urinary protein. *Eur J Biochem* **266**, 1210–1218.

Luiten G.M., Koolhaas J.M., de Boer S. and Koopmans S.J. (1985). The cortico-medial amygdala in the central nervous system organisation of agonistic behaviour. *Brain Res* **332**, 283–297.

Ma W. and Klemm W.R. (1997). Variations in equine urinary volatile compounds during the estrus cycle. *Vet Res Comm* **21**, 437–446.

Ma W., Miao Z. and Novotny M. (1998). Role of the adrenal gland and adrenal-mediated chemosignals in suppression of estrus in the house mouse: the Lee-Boot effect revisited. *Biol Reprod* **59**, 1317–1320.

Ma W., Miao Z. and Novotny M.V. (1999). Induction of estrus in grouped female mice (*Mus domesticus*) by synthetic analogues of preputial gland constituents. *Chem Senses* **23**, 289–293.

Ma W.D., Wiesler D. and Novotny M.V. (1999). Urinary volatile profiles of the deermouse (*Peromyscus maniculatus*) pertaining to gender and age. *J Chem Ecol* **25**, 417–431.

Macchi G. (1951). The ontogenetic development of the olfactory telencephalon in man. *J Comp Neurol* **95**, 245–303.

McAdam D.W. and Way J.S. (1967). Olfactory discrimination in the Giant Anteater. *Nature* **214**, 316–317.

MacCotter R.E. (1912). The connection of the vomeronasal nerves with the accessory olfactory bulb in the Opossum and other mammals. *Anat Rec* **6**, 299–318.

MacDonald D.W. and Krantz K. (1984). Behavioral, anatomical and chemical aspects of scent marking amongst Capybaras. *J Zool* **202**, 341–360.

Macintosh T.K. and Drickamer L.C. (1977). Excreted urine, bladder urine, and the delay of sexual maturation in female house mice. *Anim Behav* **25**, 999–1004.

Mackay-Sim A., Duvall D. and Graves B.M. (1985). The West Indian Manatee (*Trichechus manatus*) lacks a vomeronasal organ. *Brain Behav Evol* **27**, 186–194.

Mackay-Sim A. and Laing D. (1981). The sources of odors from stressed rats. *Physiol Behav* **27**, 511–513.

Mackay-Sim A. and Patel U. (1984). Cell-death in salamander neuroepithelium? *Exp Brain Res* **57**, 99–106.

Mackay-Sim A. and Rose J.D. (1986). Removal of vomeronasal organ impairs lordosis in female hamsters: effect is reversed by LHRH. *Neuroendocrinol* **42**, 489–493.

Maclean J., Shipley M., *et al.* (1989). Chemoanatomical organisation of the noradrenergic input to the olfactory bulb of the adult rat. *J Comp Neurol* **285**, 339–349.

MacLean P.D. (1990). *The Triune Brain in Evolution*. Plenum, New York, p. 672.

MacLeod N. and Reinhardt W. (1983). An electrophysiological study of the accessory olfactory bulb in rabbit — I. Analysis of electrical evoked potential fields. *Neuroscience* **10**, 119–129.

Macrides F., Bartke A., *et al.* (1974). Effects of exposure to vaginal odor and receptive females on plasma testosterone in male hamsters. *Neuroendocrinol* **15**, 355–364.

Macrides F., Johnson P.A. and Schneider S.P. (1977). Responses of the male golden hamster to vaginal secretion and dimethyl disulfide: attraction versus sexual behaviour. *Behav Biol* **20**, 377–386.

Madison D. (1977). Chemical communication in Amphibians and Reptiles. In: *Chemical Signals in Vertebrates* **1** (Müller-Schwarze D. and Mozell M.M., eds.). Plenum, New York, pp. 135–168.

Magert H., Cieslak A., Alkan O., Luscher B., *et al.* (1999). The golden hamster aphrodisin gene — structure, expression in parotid glands of female animals, and comparison with a similar murine gene. *J Biol Chem* **274**, 444–450.

Magert H., Hadrys T., Cieslak A., Groger A., *et al.* (1995). cDNA sequence, and expression pattern of the putative pheromone carrier aphrodisin. *Proc Natl Acad Sci* **92**, 2091–2095.

Maier W. (1980). Nasal structures in Old and New World primates. In: *Evolutionary Biology of the New World Monkeys and Continental Drift* (Ciochon A.I. and Chiarelli A.B., eds.). Karger, Basel, pp. 219–241.

Maier W. (1986). Nasal capsule and facial skeleton of a fetus of *Daubentonia Madagascariensis*. *Primate Rep* **14**, 127 abs.

Maier W. (1997). The nasopalatine duct and the nasal floor cartilages in Catarrhine primates. *Z Morph Anthrop* **81**, 289–300.

Maier W., van den Heever J. and Durand F. (1996). New therapsid specimens and the origin of the secondary hard and soft palate of mammals. *J Zool Systemat Evol Res* **34**, 9–19.

Malnic B., Hirono J., Sato T. and Buck L. (1999). Combinatorial receptor codes for odors. *Cell* **96**, 713–723.

Mancini M., Majumder D., Chatterjee B. and Roy A.K. (1989). a2μ-Globulin in modified sebaceous glands with pheromonal functions: localisation of mRNA and protein in preputial, Meibomian and perianal glands. *J Histochem Cytochem* **37**, 148–157.

Mann G. (1961). *Bulbus olfactorius accessorius* in Chiroptera. *J Comp Neurol* **116**, 135–144.

Manning C., Wakeland E. and Potts W. (1992). Communal nesting patterns in mice implicate MHC genes in kin recognition. *Nature* **360**, 581–583.

Marchese S., Pes D., Scaloni A., Carbone V. and Pelosi P. (1998). Lipocalins of boar salivary glands: binding odours and pheromones. *Europ J Biochem* **252**, 563–568.

Marchlewska-Koj A. (1981). Pregnancy block elicited by male urinary peptides in mice. *J Reprod Fertil* **61**, 221–224.

Marchlewska-Koj A. (1997). Sociogenic stress and reproductive activity in rodents. *Neurosci Biobehav Rev* **21**, 699–703.

Marchlewska-Koj A., Pochron E. and Slilwowska, A. (1990). Salivary glands and preputial glands of mice as a source of estrus-stimulating pheromone in female mice. *J Chem Ecol* **16**, 2817–2822.

Marlier L., Schaal B. and Soussignan R. (1997). Orientation responses to biological odours in the human newborn: initial pattern and postnatal plasticity. *C R Acad Sci Paris LS* **320**, 999–1005.

Martin I.G. and Beauchamp G.K. (1982). Olfactory recognition of individuals by male cavies (*Cavia aperea*). *J Chem Ecol* **8**, 1241–1250.

Marinier S.L., Alexander A.J. and Waring G.H. (1988). Flehmen behaviour in domestic horse: discrimination of conspecific odours. *Appl Anim Behav Sci* **19**, 227–237.

Martinez-Marcos A. and Halpern M. (1999a). Differential centrifugal afferents to the anterior and posterior accessory olfactory bulb. *Neuroreport* **10**, 2011–2015.

Martinez-Marcos A. and Halpern M. (1999b). Differential projections from the anterior and posterior divisions of the accessory olfactory bulb to the medial amygdala in the opossum, *Monodelphis domestica*. *Europ J Neurosci* **11**, 3789–3799.

Martinez-Marcos A., Lanuza E. and Halpern M. (1999). Organization of the ophidian amygdala: chemosensory pathways to the hypothalamus. *J Comp Neurol* **412**, 51–68.

Martinez-Marcos A., Ubeda-Banon I. and Halpern M. (2000). Cell turnover in the vomeronasal epithelium: evidence for differential migration and maturation of subclasses of vomeronasal neurons in the adult opossum. *J Neurobiol* **43**, 50–63.

Martinez-Marcos A., Ubeda-Banon I., Deng L. and Halpern M. (2000). Neurogenesis in the vomeronasal epithelium of adult rats: evidence for different mechanisms for growth and neuronal turnover. *Neurobiology* **44**, 423–435.

Martini S., Silvotti L., Shirazi A., Ryba N.J. and Tirindelli R. (2001). Co-expression of putative pheromone receptors in the sensory neurons of the vomeronasal organ. *J Neurosci* **21**, 843–848.

Maruniak J., Desjardins C. and Bronson F.A. (1975). Adaptations for urine marking in rodents: prepuce length and morphology. *J Reprod Fertil* **44**, 567–570.

Maruniak J., Wysocki C.J. and Taylor J. (1986). Mediation of male mouse urine marking and aggression by the vomeronasal organ. *Physiol Behav* **37**, 655–657.

Mason R.T. and Crews D. (1985). Female mimicry in Garter snakes. *Nature* **316**, 59–60.

Mason R.T. (1992). Reptilian pheromones. In: *Biology of the Reptilia* (Gans C. and Crews D., eds.). Chicago University Press, Vol. 18, pp. 114–228.

Massey A. and Vandenbergh J.G. (1980). Puberty delay by a urinary cue from female house mice in feral populations. *Science* **209**, 821–822.

Mateo J.M., Holmes W.G., Bell A.M. and Turner M. (1994). Sexual maturation in male prairie voles — effects of the social environment. *Physiol Behav* **56**, 299–304.

Matsuda H., Kusakabe T., Kawakami T., Takenaka T., *et al.* (1996). Coexistence of nitric-oxide synthase and neuropeptides in the mouse vomeronasal organ. *Brain Res* **712**, 35–39.

Matsuda H., Nagahara T., Tsukada M., *et al.* (1993). Distribution of galanin immunoreactive fibres in the mouse vomerronasal organ. *BioMed Res* **14**, 177–181.

Matsunami H. and Buck L. (1997). A multigene family encoding a diverse array of putative pheromone receptors in mammals. *Cell* **90**, 775–784.

Matsuoka M., Kaba H., Mori Y. and Ichikawa M. (1997). Synaptic plasticity in olfactory memory formation in female mice. *Neurorep* **8**, 2501–2504.

Matsuoka M., Mori Y. and Ichikawa M. (1998). Morphological changes of synapses induced by urinary stimulation in the hamster accessory olfactory bulb. *Synapse* **28**, 160–166.

Matsuoka M., Yokosuka M., Mori Y. and Ichikawa M. (1999). Specific expression pattern of Fos in the accessory olfactory bulb of male mice after exposure to soiled bedding of females. *J Neurosci Res* **35**, 189–195.

Matsushita F., Miyawaki A. and Mikoshiba K. (2000).Vomeroglandin/CRP-Ductin is strongly expressed in the glands associated with the mouse vomeronasal organ: identification and characterization of mouse vomeroglandin. *Biochem Biophys Res Comm* **268**, 275–281.

Matsuzaki O., Iwamaand A. and Hatanaka T. (1993). Fine-structure of the vomeronasal organ in the House Musk Shrew (*Suncus murinus*). *Zool Sci* **10**, 813–818.

McClintock M. (1983). Pheromomal regulation of the ovarian cycle: enhancement, suppression and synchrony. In: *Pheromones and Reproduction in Mammals* (Vandenberg J.G., ed.). Academic Press, New York, pp. 113–150.

McClintock T. (2000). Molecular biology of olfaction. In: *Neurobiology of Taste and Smell*, 2nd ed. (Finger T.E., Silver W.L. and Restrepo D., eds.). Wiley-Liss, New York, pp. 179–200.

McDonnell S., Diehl N., Garcia M. and Kenney R. (1989). Gonadotrophin releasing hormone (GnRH) affects precopulatory behavior in testosterone-treated geldings. *Physiol Behav* **45**, 145–149.

McFadyen D. and Addison-Wand J.L. (1999). Genomic organization of the rat alpha 2μ-globulin gene cluster. *Mamm Genome* **10**, 463–470.

Mechref Y., Ma W., Hao G. and Novotny M.V. (1999). N-linked oligosaccharides of vomeromodulin, a putative pheromone transporter in rat. *Biochem Biophys Res Comm* **255**, 451–455.

Meek L., Lee T., Rogers E. and Hernandez R. (1994). Effect of vomeronasal organ removal on behavioral estrus and mating latency in female meadow voles (*Microtus pennsylvanicus*). *Biol Reprod* **51**, 400–404.

Meisami E. and Bhatnagar K. (1998). Structure and diversity in mammalian accessory olfactory bulb. *Micros Res Tech* **43**, 476–499.

Meisami E., Mikhail L., Baim D. and Bhatnagar K.P. (1998). Human olfactory bulb: aging of glomeruli and mitral cells and a search for the accessory olfactory bulb. In: *Olfaction and Taste, XII* (Murphy C., ed.), *Ann NY Acad Sci* **855**, 708–715.

Menco B. (1997). Ultrastructural aspects of olfactory signaling. *Chem Senses* **22**, 295–311.

Mendoza A. (1986). The mouse vomeronasal glands: a light and electon microscope study study. *Chem Senses* **11**, 541–555.

Mendoza A. (1993). Morphological studies on the rodent main and accessory olfactory systems: the *regio olfactoria* and vomeronasal organ. *Anat Anz* **175**, 425–446.

Mendoza A., Krishna A., Endler J. and Kuhnel W. (1992). The olfactory region of the bat *Scotophilus heathi*. *Anat Anz* **174**, 207–211

Mendoza A.S., Kuderling I., Kuhn H.J. and Kuhnel W. (1994). The vomeronasal organ of the New-world monkey *Saguinus fuscicollis* (Callitrichidae) — a light and transmission-electron microscopic study. *Ann Anat* **176**, 217–222.

Mendoza A.S. and Kuehnel W. (1989). Light and electron microscopical studies on the vomeronasal organ of the newborn guinea pig. *Z Mikrosk–Anat Forsch* **103**, 801–806.

Mennella J.A. and Moltz H. (1988). Infanticide in the male rat: the role of the vomeronasal organ. *Physiol Behav* **42**, 303–306.

Mennella J.A. and Moltz H. (1989). Pheromonal emission by pregnant rats protects against infanticide by nulliparous conspecifics. *Physiol Behav* **46**, 591–596.

Meredith M. (1982). Stimulus access and other processes involved in nasal chemosensory function: potential substrates for neuronal and hormonal influence. In: *Olfaction and Endocrine Regulation* (Breipohl W., ed.). IRL Press, London, pp. 223–248.

Meredith M. (1983). Sensory physiology of pheromone communication. In: *Pheromones and Reproduction in Mammals* (Vandenbergh J.G., ed.). Academic Press, New York, pp. 200–252.

Meredith M. (1986). Vomeronasal organ removal before sexual experience impairs male hamster mating behavior. *Physiol Behav* **36**, 737–743.

Meredith M. (1991a). Sensory processing in the main and accessory olfactory systems: comparisons and contrasts. *J Ster Biochem Molec Biol* **39**(4b), 601–614.

Meredith M. (1991b). Vomeronasal damage, not nasopalatine duct damage, produces mating behavior deficits in male hamsters. *Chem Senses* **16**, 155–167.

Meredith M. (1994). Chronic recording of vomeronasal pump activation in awake behaving hamsters. *Physiol Behav* **56**, 345–354.

Meredith M. (1998). Vomeronasal, olfactory and hormonal convergence in the brain — cooperation or coincidence? *Ann NY Acad Sci* **855**, 349–361.

Meredith M. and Burghardt G. (1978). Electrophysiological studies of the tongue and accessory olfactory bulb in garter snakes. *Physiol Behav* **21**, 1001–1108.

Meredith M. and Fernandez-Fewell G. (1994). Vomeronasal system, LHRH, and sex behaviour. *Psychoneuroendocrinol* **19**, 657–672.

Meredith M. and Howard G. (1992). Intra-cerebroventricular LHRH relieves behavioral deficits due to vomeronasalectomy. *Brain Res Bull* **29**, 75–79.

Meredith M., Marques D.M., O‘Connell R.J. and Stern F.L. (1980). Vomeronasal pump: significance for hamster sexual behaviour. *Science* **207**, 1224–1245.

Meredith M. and O’Connell R.J. (1979). Efferent control of stimulus access to the hamster vomeronasal organ. *J Physiol* **286**, 301–316.

Meredith M. and O‘Connell R.J. (1988). HRP uptake by olfactory and vomeronasal receptor neurones: use as indicator of incomplete lesions and relevance for non-volatile chemoreception. *Chem Senses* **13**, 487–515.

Meyer D.L., Jadhao A. Bhargava S. and Kicliter E. (1996). Bulbar representation of the “water-nose” during *Xenopus* ontogeny. *Neurosci Lett* **220**, 109–112.

Meyer D.L., Jadhao A.G. and Kicliter E. (1996). Soybean agglutinin binding by primary olfactory and primary accessory olfactory projections in different frogs. *Brain Res* **722**, 222–226.

Meyer D.L., Jadhao A.G., Kiclite R., *et al.* (1997). Differential labelling of primary olfactory system sub-components by soybean agglutinin binding and NADOH-D histochemistry in the frog *Pipa*. *Brain Res* **762**, 275–280.

Mezler M., Konzelmann S., Freitag J., Rossler P. and Breer H. (1999). Expression of olfactory receptors during development in *Xenopus laevis*. *J Exp Biol* **202**, 365–376.

Miller L.R. and Gutzke W.H.N. (1999). The role of the vomeronasal organ of crotalines (Viperidae) in predator detection. *Anim Behav* **58**, 53–57.

Milton K. (1985). Urine-washing behavior in the Woolly Spider Monkey (*Brachyteles*). *Z Tierpsychol* **67**, 154–160.

Mirgall F., Breipohl W. and Bhatnagar K. (1979). Ultra-structural investigation on the cell membranes of the vomeronasal organ in the rat — a freeze-etching study. *Cell Tiss Res* **200**, 397–408.

Mitchell J.B. and Gratton A. (1992). Mesolimbic dopa release elicited by activation of AOS: a chronoamperic study. *Neurosci Lett* **140**, 81–84.

Miwa T., Moriizumi T., Sakashita H. and Kimura Y. (1993). Transection of the olfactory nerves induces expression of nerve growth factor receptor in mouse olfactory epithelium. *Neurosci Lett* **155**, 96–98.

Miyawaki A., Matsushita F., Ryo Y. and Mikoshiba K. (1994). Possible pheromone-carrier function of 2 Lipocalin proteins in the vomeronasal organ. *EMBO J* **13**, 5835–5842.

Mombaerts P. (1999). Odorant receptor genes in humans. *Curr Opin Genet Dev* **9**, 315–320.

Montagna W. (1972). Skin glands of non-human primates. *Am Zool* **12**, 109–124.

Monti-Bloch L., Diaz-Sanchez V. and Jennings-White C. (1998). Modulation of serum testosterone and autonomic function through stimulation of the male human vomeronasal organ with pregna-4, 20-diene-3,6-dione. *J Steroid Biochem Molec Biol* **65**, 237–242.

Monti-Bloch L., Jennings-White C. and Berliner D. (1998). The human vomeronasal system: a review. *Ann NY Acad Sci* **855**, 373–389.

Moore P.A., Atema J. and Gerhardt G.A. (1992). The structure of environmental odor signals: from turbulent dispersion to movement through boundary layers and mucus. In: *Chemical Signals in Vertebrates* **6** (Doty R.L. and Müller-Schwarze D., eds). Plenum, New York, pp. 79–83.

Moran D.T., Monti-Bloch L., Stensaas, L. and Berliner, D. (1994). Structure and function of the human vomeronasal organ. In: *Handbook of Olfaction and Gustation* (Doty R., eds.). M. Dekker, New York & Basel, pp. 793–820.

Moran D.T., Rowley J.C. and Jafek B.W. (1982). Electron microscopy of human olfactory epithelium reveals a new cell type: microvillar cell. *Brain Res* **253**, 39–46.

Mori K., Imamura K., Fujita K. and Obata K. (1987). Projections of two subclasses of vomeronasal nerve fibers to the accessory olfactory bulb in the rabbit. *Neuroscience* **20**, 259–278.

Mori K., von Campenhausen H. and Yoshihara Y. (2000). Zonal organization of the mammalian main and accessory olfactory systems. *Phil Trans Roy Soc: B* **355**, 1801–1812.

Morita Y. and Finger T. (1998). Differential projections of ciliated and microvillous olfactory receptor cells in the catfish, *Ictalurus punctatus*. *J Comp Neurol* **398**, 539–550.

Morofushi M., Shinohara K.F.T. and Kimura F. (2000). Positive relationship between menstrual synchrony and ability to smell 5alpha-androst-16-en-3alpha-ol. *Chem Senses* **25**, 407–411.

Moss R.L., Flynn R.E., Shi J., Shen X.M., *et al.* (1998). Electrophysiological and biochemical responses of mouse vomeronasal receptor cells to urine-derived compounds: possible mechanism of action. *Chem Senses* **23**, 483–489.

Mossing T. and Damber J.E. (1981). Rutting behaviour and androgen variation in Reindeer. *J Chem Ecol* **7**, 377–389.

Mossman C.A. and Drickamer L.C. (1996). Odor preferences of female house mice (*Mus domesticus*) in seminatural enclosures. *J Comp Psychol* **110**, 131–138.

Moulton D., Celebi G. and Fink R. (1970). Olfaction in Mammals — two aspects: proliferation of cells in the olfactory epithelium and sensitivity to odors. In: *Taste and Smell in Vertebrates* (Wolstenholme G. and Knight J., eds.). Ciba, London, pp. 227–250.

Mozell M. (1970). Evidence for a chromatographic model of olfaction. *J Gen Physiol* **56**, 46–63.

Mozell M. and Hornung D. (1984). Initial events influencing olfactory analysis. In: *Comparative Physiology of Sensory Systems* (Bolis L., *et al.*, eds.). Cambridge University Press.

Mucignat-Caretta C., Caretta A. and Baldini J (1998). Protein-bound male urinary pheromones: differential responses according to age and gender. *Chem Senses* **23**, 67–70.

Mucignat-Caretta C., Caretta A. and Cavaggioni A. (1995). Acceleration of puberty onset in female mice by male urinary proteins. *J Physiol* **486**, 517–522.

Muller F. and O'Rahilly J. (1988). The development of the human brain, including the longitudinal zoning the diencephalon at stage 15. *Anat Embryol* **179**, 55–72.

Muller J. and Marc R. (1984). Three distinct morphological classes of receptors in fish olfactory organs. *J Comp Neurol* **222**, 482–495.

Müller-Schwarze D., Silverstein R., Müller-Schwarze C., Singer A., *et al.* (1976). Response to a mamalian pheromone and its geometric isomer. *J Chem Ecol* **2**, 389–398.

Murakami S. and Arai Y. (1992). Origin of LHRH-neurons in Newts: effect of olfactory placode ablation. *Cell Tiss Res* **269**, 21–27.

Nagahara T., Matsuda H., Kadota T. and Kishida R. (1995). Development of substance-P immunoreactivity in the mouse vomeronasal organ. *Anat Embryol* **192**, 107–115.

Naito T., Saito Y., Yamamoto J., Nozaki Y., *et al.* (1998). Putative pheromone receptors related to the Ca^{2+}-sensing receptor in *Fugu. Proc Natl Acad Sci USA* **95**, 5178–5181.

Nakajima T., Sakaue M.K., Saito S., Ogawa K. and Taniguchi K. (1998). Immunohistochemical and enzyme-histochemical study on the accessory olfactory bulb of the dog. *Anat Rec* **252**, 393–402.

Nef P., Hermans-Borgmeyer I., Artieres-Pin H., Beasley L., *et al.* (1992). Spatial pattern of receptor expression in the olfactory epithelium. *Proc Natl Acad Sci* **89**, 8948–8952.

Nef S., Allaman I., Fiumelli H., De Castro E., *et al.* (1996). Olfaction in birds: differential embryonic expression of 9 putative OR genes in avian olfactory system. *Mech Dev* **55**, 65–77.

Nevison C., Barnard J., Beynon J. and Hurst J. (2000). The consequences of inbreeding for recognizing competitors. *Proc Roy Soc Lond B* **267**, 687–694.

Nevitt G. (1991). Do fish sniff? A new mechanism of olfactory sampling in pleuronectid flounders. *J Exp Biol* **157**, 1–18.

Nevo E., Bodmer M. and Heth G. (1976). Olfactory discrimination as an isolating mechanism in speciating mole rats. *Experientia* **32**, 1511–1512.

Newman S. (1999). The medial extended amygdala in male reproductive behavior — a node in the mammalian social behavior network. *Ann NY Acad Sci* **877**, 242–257.

Nishimura K., Utsumi K., Yuhara M., Fujitani Y., *et al.* (1989). Identification of puberty-accelerating pheromones in male mouse urine. *J Exp Zool* **251**, 300–305.

Nixon A., Mallet A.I., *et al.* (1988). Simultaneous quantification of five odorous steroids 16-Androstenes, in the axillary hair of men. *J Steroid Biochem* **29**, 505–510.

Nolte D., Mason J., Epple G., Aronov E. and Campbell D. (1994). Why are predator urines aversive to prey? *J Chem Ecol* **20**, 1505–1516.

Norgren R.J., Gao C., Ji Y. and Fritzsch B. (1995). Tangential migration of luteinizing hormone-releasing hormone (LHRH) neurons in the medial telencephalon in association with transient axons extending from the olfactory nerve. *Neurosci Lett* **202**, 9–12.

Northcutt R.G. and Puzdrowski R. (1988). Projections of the olfactory bulb and *Nervus terminalis* in the Silver Lamprey. *Brain Behav Evol* **2**, 96–107.

Novotny M., Jemiolo B., Harvey S., Wiesler D., *et al.* (1986). Adrenal-mediated endogenous metabolites inhibit puberty in female mice. *Science* **231**, 722–725.

Novotny M., Harvey S. and Jemiolo B. (1990). Chemistry of male dominance in the house mouse. *Experientia (Basel)* **46**, 109–113.

Novotny M., Harvey S. and Jemiolo B. (1995). Stereoselectivity in mammalian chemical communication: male mouse pheromones. *Experientia (Basel)* **51**, 738–743.

Novotny M., Jemiolo B. and Harvey S. (1990). Chemistry of rodent pheromones, molecular insights into chemical signalling in mammals. In: *Chemical Signals in Vertebrates* **5** (MacDonald D. and Natynczuk S., eds.). Oxford University Press, pp. 1–22.

Novotny M., Ma W., Zidek L. and Daev E. (1999). Recent biochemical insights into puberty acceleration, estrus induction, and puberty delay in the house mouse. In: *Advances in Chemical Signals in Vertebrates* (Johnston R.L., Müller-Schwarze D. and Sorenson P., eds.), pp. 99–116.

Novotny M., Ma W., Wiesler D. and Zidek L. (1999). Positive identification of the puberty-accelerating pheromone of the house mouse: the volatile ligands associating with the major urinary protein. *Proc Roy Soc Lond (B)* **266**, 2017–2022.

Novotny M., Jemiolo B., Wiesle D., Ma W., *et al.* (1999). A unique urinary constituent, 6-hydroxy-6-methyl-3-heptanone, is a pheromone that accelerates puberty in female mice. *Chem Biol* **6**, 377–383.

Noyes H.J. (1935). Naso-Palatine duct and Jacobson's Organ in new-born infants. *J Dent Res* **15**, 155–156.

Nyby J. (1983). Volatile and non-volatile chemosignals of female rodents: differences in hormonal regulation. In: *Chemical Signals in Vertebrates* **3** (Müller-Schwarze D. and Silverstein R.M., eds.). Plenum, New York, pp. 179–194.

Nyby J., Bigelow J., Kerchner M. and Barbehenn F. (1983). Male mouse (*Mus musculus*) ultrasonic vocalizations to female urine: why is heterosexual experience necessary? *Behav Neur Biol* **38**, 32–46.

O'Connell R.J., Singer A.G., Pfaffmann C. and Agosta W. (1979). Pheromones of hamster vaginal discharge: attraction to femtogram amounts of dimethyl disulfide and to mixtures of volatile components. *J Chem Ecol* **5**, 575–585.

O'Connell R.J. and Meredith (1984). Effects of volatile and non-volatile chemical signals on male sex behaviors mediated by the main and accessory olfactory systems. *Behav Neurosci* **98**, 1083–1093.

O'Connell R.J., Constanzo R.M. and Hildebrandt J.D. (1990). Adenylyl cyclase activation and electrophysiological responses elicited in male hamster olfactory receptor neurons by components of female pheromones. *Chem Senses* **15**, 725–740.

O'Riain M., Jarvis J.U.M. and Faulkes C. (1996). A dispersive morph in the naked mole-rat. *Nature* **380**, 619–621.

Oelschlaeger H.A. (1989). Early development of the olfactory and *N. terminalis* systems in Baleen whales. *Brain Behav Evol* **34**, 171–183.

Oelschlaeger H.A. (1989). Development of the *N. terminalis* in mammals, including toothed whales and humans. *Ann NY Acad Sci* **519**, 447–464.

Oelschlaeger H.A. and Buhl E.H. (1985). Development and rudimentation of the peripheral olfactory system in the harbor porpoise *Phacoena phacoena*, Mammalia, Cetacea. *J Morphol* **184**, 351–360.

Oelschlager H.A. (1992). Development of the olfactory and *terminalis* systems in whales and dolphins. In: *Chemical Signals in Vertebrates* **6** (Doty R.L. and Müller-Schwarze D., eds), pp. 141–147.

Ohno K., Kawasaki Y., Kubo T. and Tohyama M. (1996). Differential expression of OBP genes in rat nasal glands: implications for OBP-II as a possible pheromone transporter. *Neuroscience* **71**, 355–366.

Oikawa T., Shimamura K., Saito T. and Taniguchi K. (1993). Fine structure of the vomeronasal organ in the House Musk Shrew (*Suncus murinus*). *Exp Anim* **42**, 411–419.

Oikawa T., Shimamura K., Saito T. and Taniguchi K. (1994). Fine-structure of the vomeronasal organ in the chinchilla (*Chinchilla laniger*). *Exp Anim* **43**, 487–497.

Oikawa T., Suzuki K., Saito T.R., Tatahashi K.W., *et al.* (1998). Fine structure of three types of olfactory organs in *Xenopus laevis*. *Anat Rec* **252**, 301–310.

Ojima H. and Yamasaki T. (1988). Cholinergic innervation of the main and the accessory olfactory bulbs of the rat as revealed by a monoclonal antibody against choline acetyltransferase. *Anat Embryol* **178**, 481–488.

Okabe H., Okubo T. and Ochi Y. (1996). Expression of an epithelial membrane glycoprotein by neurons arising from the human olfactory plate through development. *Neuroscience* **72**, 579–584.

Okamoto K., Tokumitsu Y. and Kashiwayanagi M. (1996). Adenylyl cyclase activity in Turtle vomeronasal and olfactory epithelium. *Biochem Biophys Res Comm* **220**, 98–101.

Okere C., Kaba H. and T.H. (1996). Formation of an olfactory recognition memory in mice: reassessment of the role of nitric oxide. *Neuroscience* **71**, 349–354.

Olsen K.H., Grahn M., Lohm J. and Langefor A., (1998). MHC and kin discrimination in juvenile Arctic charr, *Salvelinus alpinus* (L.). *Anim Behav* **56**, 319–327.

Onoda N., Imamura K., Ariki T. and Iino M. (1981). Neocortical responses to odors in the dog. *Proc Jap Acad* **58**(B), 355–358.

Orbach J. and Kling A. (1966). Effect of sensory deprivation on onset of puberty, mating, fertility and gonadal weights in rats. *Brain Res* **73**, 141–149.

Ortmann R. (1988). Uber Sinneszellen am fetalen vomeronasalen organ der Menschen (receptor cells in the vomeronasal (Jacobson's) organ in the human fetus). *HNO* **37**, 191–197.

Osada T., Ikai A., Costanzo R.M., Matsuoka M., *et al.* (1999). Continual neurogenesis of vomeronasal neurons *in vitro*. *J Neurobiol* **40**, 226–233.

Osada T., Takezawa S., Itoh A., Arakawa H., *et al.* (1999). The distribution of sugar chains on the vomeronasal epithelium observed with an atomic force microscope. *Chem Senses* **24**, 1–6.

Osman Hill W.C. (1948). Rhinoglyphics: epithelial sculpture of the naked philtrum. *Proc Zool Soc Lond* **118**, 1–35.

Osman Hill W.C. (1972). *Evolutionary Biology of the Primates.* Academic Press, London/New York, p. 273.

Pantages E. and Dulac C. (2000). A novel family of candidate pheromone receptors in mammals. *Neuron* **28**, 835–845.

Parrott R. (1978). Courtship and copulation in pre-pubertally castrated male sheep (wethers) treated with 17α-estradiol, aromatisable androgens or DHT. *Horm Behav* **11**, 20–27.

Parsons T. (1967). Evolution of the nasal structures in the lower Tetrapods. *Am Zool* **7**, 397–413.

Parsons T. (1970). The nose and Jacobson's Organ. In: *Biology of the Reptilia* (Gans C. and Parsons T., eds.). Academic Press, London, Vol. 2B, pp. 99–191.

Parsons T. (1971). Anatomy of nasal structures from a comparative viewpoint. In: *Handbook of Sensory Physiology: Chemical Senses, Pt. 1 "Olfaction"* (Beidler L.M., ed.). Springer, Berlin, Vol. 4, pp. 1–26.

Pearce G.P. and Patterson A.M. (1992). Physical contact with the boar is required for maximum stimulation of puberty in the gilt because it allows transfer of boar pheromones and not because it induces cortisol release. *Anim Reprod Sci* **27**, 209–224.

Pearl C., Cervantes M., Chan M., Ho U., *et al.* (2000). Evidence for a mate-attracting chemosignal in the dwarf African clawed frog *Hymenochirus* spp. *Horm Behav* **38**, 67–74.

Pearson A.A. (1942). The development of the olfactory nerve, *Nervus terminalis* and vomeronasal nerve in man. *Ann ORL* **51**, 317–333.

Pelengaris S.A., Abbott D.H., Barrett J. and Moore H. (1992). Induction of estrous and ovulation in the female Grey Short-tailed Opposum *Monodelphis domestica* involves the main olfactory epithelium. In: *Chemical Signals in Vertebrates* **6** (Doty R.L. and Müller-Schwarze D., eds.). Plenum, New York, pp. 253–258.

Pelosi P. (1994). Odourant-binding proteins. *Crit Rev Biochem Molec Biol* **29**, 199–228.

Pelosi P. (1998). Odourant-binding proteins: structural aspects. *Ann NY Acad Sci* **855**, 333–348.

Perret M. (1992). Environmental and social determinants of sexual function in the male Lesser Mouse Lemur (*Microcebus murinus*). *Folia Primatol (Basel)* **59**, 1–25.

Perret M. (1995). Chemocommunication in reproductive function of Mouse Lemurs. In: *Creatures of the Dark: The Nocturnal Prosimians* (Altermann L., Doyle G. and Izard M., eds.). Plenum, New York, pp. 377–392.

Perret M. (1996). Manipulation of sex ratio at birth by urinary cues in a prosimian primate. *Behav Ecol Sociobiol* **38**, 259–266.

Perret M. and Barek S. (1991). Male influence on estrus cycles in female Woolly Opossum (*Caluromys philander*). *J Reprod Fertil* **91**, 557–566.

Perret M. and Schilling A. (1987). Role of prolactin in a pheromone-like sexual inhibition in the male Lesser Mouse Lemur. *J Endocrinol* **114**, 279–287.

Perret M. and Schilling A. (1995). Sexual responses to urinary chemosignals depend on photoperiod in a male primate. *Physiol Behav* **58**, 633–639.

Perrin T. and Rasmussen L.E.L. (1994). Chemosensory responses of female Asian elephants (*Elephas maximus*) to cyclohexanone. *J Chem Ecol* **20**, 2953–2958.

Petrulis A. and Johnston R. (1999). Lesions centered on the medial amygdala impair scent-marking and sex-odor recognition, but spare discrimination of individual odors in female golden hamsters. *Behav Neurosci* **113**, 345–357.

Petrulis A., Peng M. and Johnston R.E. (1999). Effects of vomeronasal organ removal on individual odor discrimination, sex-odor preference, and scent marking by female hamsters. *Physiol Behav* **66**, 73–83.

Petti M.A., Matheson S.F. and Burd G.D. (1999). Differential antigen expression during metamorphosis in the tripartite olfactory system of the African clawed frog, *Xenopus laevis*. *Cell Tiss Res* **297**, 383–396.

Pfeiffer C. and Johnston R.E. (1994). Hormonal and behavioral responses of male hamsters to females and female odors: roles of olfaction, the vomeronasal system, and sexual experience. *Physiol Behav* **55**, 129–138.

Pfeiffer S. (1985). Flehmen and dominance among captive adult female Scimitar-horned Oryx (*Oryx dammah*). *J Mammal*, 160–163.

Phillips I.R. (1976). Embryology of Common Marmoset (*Callithrix jacchus*). *Adv Anat Emb Cell Biol* **52**, 1–47.

Phillips M.S., Ho B. and Linner J. (1982). Ultrastructural localization of LHRH immunoreactive synapses in the hamster accessory olfactory bulb. *Brain Res* **246**, 193–204.

Pieper D., Newman S.W., Lobocki C.A. and Gogola G. (1989). Bilateral transection of LOT but not removal of the vomeronasal organ inhibits short-day-induced testicular regression in hamster. *Brain Res* **485**, 382–390.

Pieper D. and Newman S. (1999). Neural pathway from the olfactory bulbs regulating tonic gonadotropin secretion. *Neurosci Bio Behav Revs* **23**, 555–562.

Planel H. (1953). Etudes sur la physiologie de l'Organe de Jacobson. *Arch Anat Histol Embryol* **36**, 199–205.

Plant T.M., Gay V.L., Marshall G.R. and Arslan M. (1989). Puberty in monkeys is triggered by chemical stimulation of hypothalamus. *Proc Natl Acad Sci* **86**, 2506–2510.

Plendl J. and Schmahl W. (1988). *Dolichos biflorus* agglutinin: a marker of the developing olfactory system in a NMRI-mouse strain. *Anat Embryol* **177**, 439–444.

Pocock R. (1916). On the external characters of the Mongooses. *Proc Zool Soc Lond* **1916**, 349–374.

Pomeroy S.L., LaMantia A.S. and Purves D. (1990). Post-natal construction of neural circuitry in the mouse olfactory bulb. *J Neurosci* **10**, 1952–1966.

Poole J. (1987). Rutting behavior in African elephants — the phenomenon of musth. *Behaviour* **102**, 283–316.

Poran N.S., Vandoros A. and Halpern M. (1993). Nuzzling in the Gray Short-tailed Opossum, I: Delivery of odors to vomeronasal organ. *Physiol Behav* **53**, 959–967.

Poran N. (1998). Vomeronasal organ and its associated structures in the opossum *Monodelphis domestica. Micros Res Techn* **43**, 500–510.

Porteros A., Areval O.R., Crespo C., Garciaojeda E., *et al.* (1995). Calbindin D-28K immunoreactivity in the rat accessory olfactory bulb. *Brain Res* **689**, 93–100.

Porteros A., Brinon J., Crespo C., Okazakik H., *et al.* (1996). Neurocalcin immunoreactivity in the rat accessory olfactory bulb. *Brain Res* **729**, 82–89.

Prescott R. (1977). A mechanism for the presentation of chemical stimuli to the vomeronasal organ in the cat. *J Anat* **123**, 244–245 (abs).

Preti G., *et al.* (1990). Human axillary odors: their communicative function, structure and origin. *Proc Am Chem Soc* **200**, 12 (abs).

Preti G., Spielman A., Zeng X.-N. and Leyden J.J. (1995). The characteristic female axillary odors and their precursor proteins: qualitative compraison to males. *Chem Senses* **20**, 760 (abs).

Preti G. and Wysocki C.J. (1999). Human pheromones: releasers or primers, fact or myth? In: *Advances in Chemical Signals in Vertebrates* (Johnston R.E., Müller-Schwarze D. and Sorenson P., eds.). Kluwer Academic Press/Plenum, New York, pp. 315–332.

Preti G., Zelson P.R., Kostelc J.G. and Huggins G.R. (1980). Cyclic variations in salivary volatiles. *J Dent Res* **59**(suppl. A), 356.

Probst B. (1990). Female urinary chemosignals stimulate scent marking behaviour in male Mongolian Gerbils. In: *Chemical Signals in Vertebrates* **5** (MacDonald D., Müller-Schwarze D. and Natynczuk S.E., eds.). University Press, Oxford, pp. 213–216.

Purvis K. and Haynes N. (1978). Odours of female rat urine on plasma testosterone in male rats. *J Reprod Fertil* **53**, 63–65.

Pyatkina G.A. (1987). Development of the receptor cells of the olfactory and vomeronasal organs in man. *Cell Differ* **20**(Suppl.), 91S.

Qasba P. and Reed R. (1998). Tissue and zonal-specific expression of an olfactory receptor transgene. *J Neurosci* **18**, 227–236.

Quaglino E., Giustetto M., Panzanelli P., Cantino D., *et al.* (1999). Immunocytochemical localization of glutamate and gamma-aminobutyric acid in the accessory olfactory bulb of the rat. *J Comp Neurol* **408**, 61–72.

Quay W.B. (1972). Integument and the environment: glandular composition, function and evolution. *Am Zool* **12**, 95–108.

Quay W.B. (1977). Structure and function of scent glands. In: *Chemical Signals in Vertebrates* **1** (Müller-Schwarze D. and Mozell M.M., eds.). Plenum, New York, pp. 1–16.

Quinton H.W., Grant W., Thrasivoulou C.Q. and Besser G.M. (1997). Gonadotropin-releasing hormone immunoreactivity in the nasal epithelia of adults with Kallmann's syndrome and isolated hypogonadotropic hypogonadism, in the early midtrimester human fetus. *J Clin Endocrinol Metab* **82**, 309–314.

Raisman G. (1972). An experimental study of the projection of the amygdala to the accessory olfactory bulb and its relationship to the concept of a dual olfactory system. *Exp Brain Res* **14**, 395–404.

Rajendren G. and Dominic J. (1986). Effect of bilateral transection of the lateral olfactory tract on the male-induced implantation failure (Bruce effect) in mice. *Physiol Behav* **36**, 587–590.

Rajendren G., Dudley C. and Moss R. (1993). Influence of male rats on the LHRH neuronal system in the female: role of vomeronasal organ. *Neuroendocrinology* **57**, 898–906.

Rajendren G. and Moss R. (1994). Vomeronasal organ-mediated induction of *fos* in the CNS pathways of repetitively mated female rats. *Brain Res Bull* **34**, 53–59.

Rama Krishna N.S., Getchell M.L. and Getchell T.V. (1994). Expression of the putative pheromone and odorant transporter vomeromodulin mRNA and protein in nasal chemosensory mucosae. *J Neurosci Res* **39**, 243–259.

Rama-Krishna N.S., Getchell M., Margolis F. and Getchell T.V. (1995). Differential expression of vomeromodulin and odorant-binding protein, putative pheromone and odorant transporters, in the developing rat nasal chemosensory mucosae. *J Neurosci Res* **1**, 54–71.

Ranson E. and Beach F.A. (1985). Effects of testosterone on ontogeny of urinary behavior in male and female dogs. *Horm Behav* **19**, 36–51.

Rasia A., Londero R. and Achaval M. (1999). Effects of gonadal hormones on the morphology of neurons from the medial amygdaloid nucleus of rats. *Brain Res Bull* **48**, 173–183.

Rasmussen L.E.L. and Hultgren B. (1990). Gross and microscopic anatomy of vomeronasal organ in the Asian Elephant (*Elephas maximus*). In: *Chemical Signals in Vertebrates* **5** (MacDonald D., Müller-Schwarze D. and Natynczuk S.E., eds.). Oxford University Press, pp. 154–161.

Rasmussen L.E.L., Lee T.D., Daves G. and Schmidt M.J. (1993). Female-to-male sex-pheromones of low volatility in the Asian Elephant, *Elephas maximus*. *J Chem Ecol* **19**, 2115–2128.

Rasmussen L.E.L., Lee T.D., Zhang A., Roelofs W.L., *et al.* (1997). Purification, identification, concentration and bioactivity of (Z)-7-dodecen-1-yl acetate: sex pheromone of the female Asian Elephant, *Elephas maximus*. *Chem Senses* **22**, 417–438.

Rasmussen L.E.L. and Munger B.L. (1996). The sensorineural specializations of the trunk tip (finger) of the Asian elephant, *Elephas maximus*. *Anat Rec* **246**, 127–134.

Rasmussen L.E.L. and Perrin T. (1999). Physiological correlates of musth: lipid metabolites and chemical composition of exudates. *Physiol Behav* **67**, 539–549.

Rasmussen L.E.L., Schmidt M.J. and Daves G. (1982). Asian bull elephants: Flehmen-like responses to extractable components of female elephant estrus urine. *Science* **217**, 159–162.

Rasmussen L.E.L. and Schulte B. (1998). Chemical signals in the reproduction of Asian (*Elephas maximus*) and African (*Loxodonta africana*) elephants. *Anim Reprod Sci* **53**, 19–34.

Rawleigh J.M., Kemble E.D. and Ostrem J. (1993) Differential effects of prior dominance or subordination experience on conspecific odor preferences in mice. *Physiol Behav* **54**, 35–39.

Read E.A. (1908). A contribution toward the knowledge of the olfactory apparatus in dog, cat and man. *Am J Anat* **8**, 17–47.

Reger R.L., Gerall A.A., *et al.* (1987). LHRH neuronal system in the accessory olfactory bulb of the prairie vole, *Microtus ochrogaster*. *Neurosci Abs* **13**, 993.

Regier F. and Goodwin M. (1977). On the chemical and environmental modulation of pheromone release from vertebrate scent marks. In: *Chemical Signals in Vertebrates* **1** (Müller-Schwarze D. and Mozell M.M., eds.), pp. 115–134.

Rehorek S.J., Firth B.T. and Hutchinson M.N. (2000). The structure of the nasal chemosensory system in squamate reptiles, 2: Lubricatory capacity of the vomeronasal organ. *J Biosci* **25**, 181–190.

Rehorek S., Hillenius W., Quan W., Halpern M., *et al.* (2000). Passage of Harderian gland secretions to the vomeronasal organ of *Thamnophis sirtalis* (Colubridae). *Can J Zool* **78**, 1284–1288.

Reilly J., Vowels B., Leyden J., Sondheimer S., *et al.* (1996). Quantitative comparison of female axillary secretions as a function of the menstrual cycle phase. *Chem Senses* **21**, 661–662.

Reinhardt W., MacLeod N., Ladewig J. and Ellendorff F. (1983). An electrophysiological study of the accessory olfactory bulb in the rabbit — II. *Neuroscience* **10**, 131–139.

Ressler K.J., Sullivan S.L. and Buck L. (1993). Zonal organisation of odorant receptor gene expression in the olfactory epithelium. *Cell* **73**, 597–609.

Rettori V., Belova N., Dees W.L., Nyberg C.L., *et al.* (1993). Role of nitric oxide in the control of luteinizing hormone-releasing hormone release *in vivo* and *in vitro*. *Proc Natl Acad Sci* **90**, 10130–10134.

Rich T.J. and Hurst J.L. (1999). The competing countermarks hypothesis: reliable assessment of competitive ability by potential mates. *Anim Behav* **58**, 1027–1037.

Risser J.M. and Slotnick B.M. (1987). Nipple attachment and survival in neonatal olfactory bulbectomized rats. *Physiol Behav* **40**, 545–550.

Ritchie J. (1944). *Thomson's Outlines of Zoology*. Oxford University Press: H. Milford London, p. 851.

Rivard G. and Klemm W. (1989a). Sample contact required for complete bull response to oestrus pheromone in cattle. In: *Chemical Signals in Vertebrates* **5** (MacDonald D.W., *et al.*, eds.). Oxford University Press, pp. 627–633.

Rivard G. and Klemm W.R. (1989b). Two body fluids containing bovine estrus pheromone(s). *Chem Senses* **14**, 273–280.

Robertson D., Benyon R.L. and Evershed R. (1993). Extraction, characterization and binding analysis of two pheromonally active ligands associated with major urinary proteins of house mouse (*Mus musculus*). *J Chem Ecol* **19**, 1405–1416.

Rodolfo-Masera T. (1943). Sur l'esistenza di un particolare organo olfativo nel setto nasale della cavia e di altri roditori. *Arch Ital Anat Embryol* **48**, 157–212.

Rodriguez I., Feinstein P. and Mombaerts P. (1999). Variable patterns of axonal projections of sensory neurons in the mouse vomeronasal system. *Cell* **97**, 199–208.

Rodriguez I., Greer C., Mok M. and Mombaerts P. (2000). A putative pheromone receptor gene expressed in human olfactory mucosa. *Nature Genet* **26**, 18–19.

Rodriquez I., Punta K.D., Rothman A., Ishii I. and Mombaerts P. (2002). Multiple new and isolated families within the mouse superfamily of V1r vomeronasal receptors. *Nature Neurosci* **5**, 134–140.

Roland R., Halpern M., *et al.* (1995). Effects of vomeronasal-axotomy on the anatomy of the accessory olfactory bulb and on nuzzling behavior in the Brazilian Short-tailed Opossum *Monodelphis domestica*. *Chem Senses* **19**, 544–545.

Rollmann S., Houck L. and Feldhoff R. (1999). Proteinaceous pheromone affecting female receptivity in a terrestrial salamander. *Science* **285**, 1907–1909.

Ronnekliev O. and Resko J.A. (1990). Ontogeny of GnRH-containing neurons in early fetal development of rhesus macaques. *Endocrinology* **126**, 498–511.

Roos J., Roos M., Schaeffer C. and Aron C. (1988). Sexual differences in the development of accessory olfactory bulbs in the rat. *J Comp Neurol* **270**, 121–131.

Roos J., Roos M., Schaeffer C. and Aron C. (1989). Prepubescent hormonal control of the development of accessory olfactory bulbs in the male rat. *Dev Brain Res* **47**, 309–312.

Rose F., Dotman R. and Wraver G. (1969). Electrophoresis of Tortoise (*Gopherus*) chin-gland extracts. *Comp Biochem Physiol* **29**, 847–851.

Rosen S., Shelesnyak M. and Zacharias L. (1940). Naso-genital relationship, II: Pseudopregnancy following extirpation of the sphenopalatine ganglion in the rat. *Endocrinology* **27**, 463–468.

Rosenblatt J.S. (1983). Olfaction mediates developmental transition in the altricial newborn of selected species of mammals. *Dev Psychobiol* **16**, 347–375.

Rossler P., Mezler M. and Breer H. (1998). Two olfactory marker proteins in *Xenopus laevis*. *J Comp Neurol* **395**, 273–280 .

Rossler P., Kroner C.K.J., Lobel D., Breer H., *et al.* (2000). Cyclic adenosine monophosphate signaling in the rat vomeronasal organ: role of an adenylyl cyclase type VI. *Chem Senses* **25**, 313–322.

Russell E. (1984). Social behaviour and organisation of marsupials. *Mammal Rev* **14**, 101–154.

Ryba N. and Tirindelli R. (1996). The G-protein subunit g-8 is expressed the developing axons of olfactory and vomeronasal neurons. *Europ J Neurosci* **8**, 2388–2398.

Ryba N. and Tirindelli R. (1997). A new multigene family of putative pheromone receptors. *Neuron* **19**, 371–379.

Rylands A. (1985). Tree-gouging and scent marking by marmosets. *Anim Behav* **33**, 1365–1367.

Sachs B.D. (1997). Erection evoked in male rats by airborne scent from estrous females. *Physiol Behav* **62**, 921–924.

Saint Girons H. (1976). Histological study of nasal organs of *Crocodylus niloticus* and Caiman — Crocodylidae. *Zoomorphol* **84**, 301–318.

Saito H., Mimmack M., Keverne E.B., Kishimoto J. and Emson P. (1998). Isolation of mouse vomeronasal receptor genes and their co-localization with specific G-protein messenger RNAs. *Molec Brain Res* **60**, 215–227.

Saito S. and Taniguchi K. (2000). Expression patterns of glyco-conjugates in the three distinctive olfactory pathways of the clawed frog, *Xenopus laevis*. *J Vet Med Sci* **62**, 153–159.

Saito T.R. and Mennella J. (1986, not seen). A simple operation method on removal of the vomeronasal organ of the rat. *Jik Dobutsu* **35**, 527–529.

Saito T.R. and Moltz H. (1986). Sexual behavior in the female rat following removal of the vomeronasal organ. *Physiol Behav* **38**, 81–87.

Saito T.R., Kamata K., Nakamura M. and Inaba M. (1988). Maternal behaviour in virgin female rats following removal of the vomeronasal organ. *Zool Sci* **5**, 1141–1144.

Salamon M. (1996). Olfactory communication in Australian marsupials. In: *Comparisons of Marsupial and Placental Behaviour* (Croft D.B., *et al.*, eds.). Furth, Filander, pp. 46–79.

Salazar I., Barber P.C. and Cifuentes J.M. (1992). Anatomical and immunohistological demonstration of the primary neural connections of the vomeronasal organ in the dog. *Anat Rec* **233**, 309–313.

Salazar I., Cifuentes J., Quinteiro P. and Caballero T. (1994). Structural, morphometric, and immunohistological study of the accessory olfactory bulb in the dog. *Anat Rec* **240**, 277–285.

Salazar I., Lombardero M., Sanchez-Quinteiro P., Roel P., *et al*.,. (1998). Origin and regional distribution of the arterial vessels of the vomeronasal organ in the sheep: a methodological investigation with SEM and cutting-grinding techniques. *Ann Anat* **180**, 181–187.

Salazar I., Quinteiro P. and Cifuentes J. (1995). Comparative anatomy of the vomeronasal cartilage in mammals — Mink, Cat, Dog, Pig, Cow and Horse. *Ann Anat* **177**, 475–481.

Salazar I., Quinteiro P. and Cifuentes J. (1997). The soft-tissue components of the vomeronasal organ in pigs, cows and horses. *Anat Histol Embryol* **26**, 179–186.

Salazar I., Quinteiro P., Cifuentes J., Fernandez P. and Lombardero M. (1997). Distribution of the arterial supply to the vomeronasal organ in the cat. *Anat Rec* **247**, 129–136.

Salazar I. and Quinteiro P. (1998). Supporting tissue and vasculature of the vomeronasal organ: the rat as a model. *Microsc Res Tech* **41**, 492–505.

Salazar I., Quinteiro P., Cifuentes J.M., and Lombardero M. (1998). The accessory olfactory bulb of the mink, *Mustela vison*: a morphological and lectin histochemical study. *J Vet Med C* **27**, 297–300.

Salazar I., Sanchez-Quinteiro P., Lombardero M. and Cifuentes J. (2000). A descriptive and comparative lectin histochemical study of the vomeronasal system in pigs and sheep. *J Anat* **196**, 15–22.

Salier J.-P. (2000). Chromosomal location, exon/intron organization and evolution of lipocalin genes. *Biochim Biophys Acta* **1482**, 25–34.

Saltiere-Imerly R.B., Young B.J. and Capozza M.A. (1989). Estrogen differentially regulates neuropeptide gene expression in a sexually dimorphic olfactory pathway. *Proc Natl Acad Sci* **86**, 4766–4770.

Sanchez-Barcelo E., Mediavilla M.D., Sanchez-Criado J.E., Cos S., *et al.* (1985). Antigonadal actions of olfaction and light deprivation: effects of blindness and main olfactory bulb — deafferentation and transection of vomeronasal nerve or bulbectomy. *J Pineal Res* **2**, 177–190.

Sanchez Criado J.E. (1982). Involvement of the vomeronasal system in the reproductive physiology of the rat. In: *Olfaction and Endocrine Regulation* (Breipohl W., ed.). IRL Press, London, pp. 209–222.

Sasaki K., Okamoto K., Inamura K., Tokumitsu Y., *et al.* (1999). Inositol-1,4,5-trisphosphate accumulation induced by urinary pheromones in female rat vomeronasal epithelium. *Brain Res* **823**, 161–168.

Saskena S. and Chandra G. (1980). Gross, histological and certain histochemical observations on the vomeronasal organ of Buffalo (*Buffalo bubalis*). *Ind J Anim Health* **19**, 99–104.

Savage R.J.G. and Long M.R. (1986). *Mammal Evolution*. British Museum (Natural History), London, p. 264.

Sawyer L. (1987). One fold among many. *Nature* **327**, 659.

Sawyer T., Miller K.V. and Marchinton R.L. (1993). Patterns of urination and rub-urination in female White-tailed Deer. *J Mammal* **74**, 477–479.

Scalfari F., Castagna M., Fattori B., Andreini I., *et al.* (1997). Expression of a lipocalin in human nasal mucosa. *Comp Biochem Physiol B* **118**, 819–824.

Scalia F. and Winans S. (1975). The differential projections of the olfactory bulb and accessory olfactory bulb in mammals. *J Comp Neurol* **161**, 31–56.

Schaal B., Marlier L. and Soussignan R. (1998). Olfactory function in the human fetus: evidence from selective neonatal responsiveness to the odor of amniotic fluid. *Behav Neurosci* **112**, 1438–1449.

Schaal B., Orgeur P. and Arnould C. (1995). Olfactory preferences in newborn lambs — possible influence of prenatal experience. *Behaviour* **132**, 351–365.

Schaal B., Orgeur P. and Rognon C. (1995). Odor sensing in the human fetus: anatomical, functional and chemo-ecological bases. In: *Prenatal Development: Psychobiological Perspectives* (Krasnegor N.A., Fifer W.A. and Smotherman W.P., eds.). L. Erlbaum Ass, Hillsdale, N.J.

Schaeffer J.P. (1910). The lateral wall of the *cavum nasi* in man with especial reference to the various developmental stages. *J Morphol* **21**, 613–707.

Schaefer J. (1940). *die Hautdrüsen organe der Säugetier*. Urban u. Schwarzenberg, Berlin & Vienna, p. 452.

Schellinck H., Smyth C., Brown R. and Wilkinson M. (1993). Odor-induced sexual maturation and expression of c-fos in the olfactory system of juvenile female mice. *Dev Brain Res* **74**, 138–141.

Schellinck H.M., West A.M. and Brown R.E. (1992). Rats can discriminate between the urine odors of genetically identical mice maintained on different diets. *Physiol Behav* **51**, 1079–1082.

Schild D. and Restrepo D. (1998). Transduction mechanisms in vertebrate olfactory receptor cells. *Physiol Revs* **78**, 429–466.

Schilling A. (1970). L'Organe de Jacobson du lémurien Malgache, *Microcebus murinus* (Miller, 1777). *Mém Mus Natl d'Hist Nat (Paris)* **61A**, 1–203.

Schilling A. (1974). A study of marking behaviour in *Lemur catta.* In: *Prosimian Biology* (Martin R.D., *et al.*, eds.). Duckworth, London, pp. 347–363.

Schilling A. (1990). Communications par signaux chimiques chez les prosimiens. In: *Primates-Recherches Actuelles* (Roeder J.J. and Anderson J.R., eds.). Masson et Cie, Paris, pp. 121–143.

Schilling A., Serviere J., Gendrot G. and Perret M. (1990). Vomeronasal activation by urine in the primate *Microcebus murinus*: a 2DG study. *Exp Brain Res* **81**, 609–618.

Schmale H., Holtgreve-Grez H. and Christiansen H. (1990). Possible role for salivary gland protein in taste reception indicated by homology to lipophillic-ligand carrier proteins. *Nature* **343**, 366–369.

Schmidt A. and Roth G. (1990). Central olfactory and vomeronasal pathways in salamanders. *J Hirnforsch* **31**, 543–553.

Schmidt A., Naujoks-Manteuffel C. and Roth G. (1988). Olfactory and vomeronasal projections and the pathway of the *Nervus terminalis* in ten species of salamanders — a whole mount study employing the horseradish-peroxidase technique. *Cell Tissue Res* **251**, 45–50.

Schmidt A. and Wake M. (1990). Olfactory and vomeronasal systems of caecilians (*Gymniophiona*). *J Morphol* **205**, 255–268.

Schmidt U. and Schmidt C. (1986). Influence of vomeronasal organ on olfactory guided behaviour in suckling mice. *Z Saugetierk* **51**, 86–90.

Schneider K. (1930–35). Das Flehmen: i–v., *Zool Gart* **3–5**, 183–198, 200–226.

Schofield P.R. (1988). Carrier-bound odorant delivery to olfactory receptors. *Trends Neurosci* **11**, 471–472.

Schwanzel-Fukuda M., Fadem B., Garcia M. and Pfaff D. (1988). Immunocytochemical localization of LHRH in the brain and *Nervus terminalis* of the adult and early neonatal Gray Short-Tailed opossum (*Monodelphis domestica*). *J Comp Neurol* **276**, 44–60.

Schwarting G. and Crandall J. (1991). Subsets of olfactory and vomeronasal sensory cells and axons revealed by MC.Ab to carbohydrate antigens. *Brain Res* **547**, 239–248.

Schwanzel-Fukuda M., Crossin K. Pfaff D.W., Bouloux P.M., *et al.* (1996). Migration of luteinizing hormone-releasing hormone neurons in early human embryos. *J Comp Neurol* **366**, 547–557.

Schwanzel-Fukuda M., Reinhard G.R., Abraham S., Crossin K.L., *et al.* (1994). Antibody to neural cell adhesion molecule can disrupt the migration of luteinizing hormone-releasing hormone neurons into the mouse brain. *J Comp Neurol* **342**, 174–185.

Schwarting G.A., Drinkwater D. and Crandall J.E. (1994). A unique neuronal glycolipid defines rostrocaudal compartmentalization in the accessory olfactory system of rats. *Brain Res* **78**, 191–200.

Schwenk K. (1985). Occurrence, distribution, and functional significance of taste buds in lizards. *Copeia*, 91–101.

Schwenk K. (1993). The evolution of chemoreception in squamate reptiles: a phylogenetic approach. *Brain Behav Evol* **41**, 124–137.

Schwenk K. (1994). Why snakes have forked tongues. *Science* **263**, 1573–1577.

Segovia S., Paniagua R., Nistal M. and Guillamon A.. (1984). Effects of post-pubertal gonadectomy on the neurosensorial epithelium of the vomeronasal organ in the rat. *Dev Brain Res* **316**, 289–291.

Segovia S. and Guillamon A. (1993). Sex dimorphism in the vomeronasal pathway and sex differences in reproductive behaviours. *Brain Res Rev* **18**, 51–74.

Segovia S., Guillamon A., del Cerro M., Ortega E., *et al.* (1999). The development of brain sex differences: a multisignaling process. *Behav Brain Res* **105**, 69–80.

Seitz E. (1969). Die bedeutung geruchlicher orientierung beim Plumplori (*Nycticebus coucang*). *Z f Tierpsychol* **26**, 73–103.

Shair H., Masmela J. and Hofer M. (1998). The influence of olfaction on potentiation and inhibition of ultrasonic vocalization of rat pups. *Physiol Behav* **65**, 769–772.

Shapiro L. and Halpern M. (1995). Lectin histochemical identification of carbohydrate moieties in opossum chemosensory systems during development, with special emphasis on VVA-identified subdivisions in the accessory olfactory bulb. *J Morphol* **224**, 331–349.

Shapiro L., Roland R.M., Li C.S. and Halpern M. (1996) Vomeronasal system involvement in response to conspecific odors in adult male opossums, *Monodelphis domestica. Behav Brain Res* **77**, 101–113.

Sharon D., Glusman G., Pilpel Y., Horn-Saban S. and Lancet D. (1998). Genome dynamics, evolution, and protein modelling in the olfactory gene superfamily. *Ann NY Acad Sci* **855**, 182–193.

Sharon D., Glusman G., Pilpel Y., Khen M., *et al.* (1999). Primate evolution of an olfactory receptor cluster; diversification by gene conversion and recent emergence of pseudogenes. *Genomics* **61**, 24–36.

Shaw P., Held W. and Hastie N. (1983). The gene family for major urinary proteins, expression in several secretory tissues of the mouse. *Cell* **32**, 755–761.

Shelley W., Hurley H. and Nichols A. (1953). Axillary odor, an experimental study of the role of bacteria, apocrine sweat, and deodorants. *Arch Dermatol* **68**, 430–446.

Shibuya T. and Tsuaki H.T. (1988). Induced wave responses of the accessory olfactory bulb to odorants in two species of turtle. *Comp Biochem Physiol* **91**, 377–386.

Shinohara K., Morofushi M., Funabashi T., Mitsushima D., *et al.* (2000). Effects of 5alpha-androst-16-en-3alpha-ol on the pulsatile secretion of luteinizing hormone in human females. *Chem Senses* **25**, 465–467.

Shipley M. and Ennis M. (1996). Functional organisation of the olfactory system. *J Neurobiol* **30**, 123–176.

Shirley S.G., Polak E., Edwards D., Wood M., *et al.* (1987). The effect of ConA on the rat electro-olfactogram at various concentrations. *Biochem J* **245**, 185–189.

Shnayder L., Schwanzel-Fukuda M. and Halpern M. (1993). Differential OMP expression in opossum accessory olfactory bulb. *Neuroreport* **5**, 193–196.

Simerly R. (1990). Hormonal control of neuropeptide gene expression in sexually dimorphic olfactory pathways. *Trends Neurosci* **13**, 104–110.

Singer A.G., *et al.* (1990). Aphrodisin: pheromone or transducer? *Chem Senses* **15**, 199–204.

Singer A.G., Beauchamp G. and Yamazaki K. (1997). Volatile signals of the major histocompatibility complex in male mouse urine. *Proc Natl Acad Sci* **94**, 2210–2214.

Singer A.G., Clancy A.N. and Macrides F. (1989). Conspecific and heterospecific proteins related to aphrodisin lack aphrodisiac activity in male hamsters. *Chem Senses* **14**, 563–576.

Singh P. and Hofer M. (1978). Oxytocin reinstates maternal olfactory cues for nipple orientation and attachment in rat pups. *Physiol Behav* **20**, 385–389.

Sipos M., Alterman L., Perry B., Nyby J., *et al.* (1995). An ephemeral pheromone of female house mice — degradation by oxidation. *Anim Behav* **50**, 113–120.

Sipos M.L., Wysocki C.J., Nyby J.G., Wysocki L., *et al.* (1995) An ephemeral pheromone of female house mice: perception via the main and accessory olfactory systems. *Physiol Behav* **58**, 529–534.

Skeen L.C. and Hall W. (1977). Efferent projection of the main and the accessory olfactory bulb in tree Shrew (*Tupaia glis*). *J Comp Neurol* **172**, 1–36.

Smith A.B., Belcher A.M., Epple G., Jurs P.C., *et al.* (1985) Computerized pattern recognition: a new technique for the analysis of chemical communication. *Science* **228**, 175–177.

Smith B.A. and Block M.L. (1991). Male saliva cues and female social choice in Mongolian gerbils. *Physiol Behav* **50**, 379–384.

Smith C.G. (1935). The change in volume of the olfactory and accessory olfactory bulbs of the albino rat during post-natal life. *J Comp Neurol* **61**, 477–508.

Smith T.D., Siegel M.I., Mooney M.P., Burdi A.R., *et al.* (1996). Vomeronasal organ growth and development in normal, and cleft-lip and palate, human fetuses. *Cleft Palate-Craniofac J* **33**, 385–394.

Smith T.D., Siegel M.I., Mooney M.P., Burdi A.R., *et al.* (1997). Prenatal growth of the human vomeronasal organ. *Anat Rec* **248**, 447–455.

Smith T.D., Siegel M.I., Burrows A.M., Mooney M.P., *et al.* (1998). Searching for the vomeronasal organ of adult humans: preliminary findings on location, structure and size. *Micros Res Tech* **41**, 483–491.

Smith T.D., Siegel M.I., Burrows A.M., Mooney M.P., *et al.* (1999). Histological changes in the fetal human vomeronasal epithelium during volumetric growth of the vomeronasal organ. In: *Advances in Chemical Signals in Vertebrates* (Johnston R.E., Müller-Schwarze D. and Sorenson P., eds.). Plenum, New York, pp. 583–592.

Smith T.E. and Abbott D.H. (1998). Behavioral discrimination between circumgenital odor from peri-ovulatory dominant and anovulatory female common marmosets (*Callithrix jacchus*). *Am J Primatol* **46**, 265–284.

Smith T.E., Abbott D.H., Tomlinson A. and Mlotkiewicz J. (1997). Differential display of investigative behavior permits discrimination of scent signatures from familiar and unfamiliar socially dominant female marmoset monkeys (*Callithrix jacchus*). *J Chem Ecol* **23**, 2523–2546.

Smith T.E., Faulkes C.G. and Abbott D.H. (1997). Combined olfactory contact with the parent colony and direct contact with nonbreeding animals does not maintain suppression of ovulation in female naked mole-rats (*Heterocephalus glaber*). *Horm Behav* **31**, 277–288.

Smithson K.G., Weiss M.L. and Hatton G.I. (1992). Supraoptic nucleus afferents from the accessory olfactory bulb: evidence from anterograde and retrograde tract-tracing in the rat. *Brain Res Bull* **29**, 209–220.

Smotherman W.P. and Robinson S.R. (1992). Habituation in the rat fetus. *Q J Exp Psychol B* **44**, 215–230.

Sobel N., Khan R., Siltman A., Sullivan E., *et al.* (1999). The world smells different to each nostril. *Nature* **402**, 35.

Sobel N., Prabhakaran V., Hartley C., Desmond J., *et al.* (1999). Blind smell: brain activation induced by an undetected air-borne chemical. *Brain* **122**, 209–217.

Sœmmerring (1809). *Abbildung der menslichen Organe des Geruches,* Frankfurt (cited by — v. Navratil, 1926).

Soler M. and Suburo A.M. (1998). Innervation of blood vessels in the vomeronasal complex of the rat. *Brain Res* **811**, 47–56.

Sorensen P.W. (1996). Biological responsiveness to pheromones provides fundamantal and unique insight into olfactory function. *Chem Senses* **21**, 245–256.

Sorensen P. and Caprio J. (1998). Chemoreception,. In: *The Physiology of Fishes*, 2nd ed. (Evans D.H., ed.). CRC Press, Boca Raton, pp. 373–410.

Sorensen P., Christensen T. and Stacey N. (1998). Discrimination of pheromonal cues in fish: emerging parallels with insects. *Curr Opin Neurobiol* **8**, 458–467.

Sorensen P.W., Hara T.J., Stacey N.E. and Goetz F. (1988). F-prostaglandins function as potent olfactory stimulants that comprise the post-ovulatory female sex pheromone in goldfish. *Biol Reprod* **39**, 1039–1050.

Sorensen P.W., Hara T.J., Stacey N.E. and Dulka J.G. (1990). Extreme olfactory specificity of male goldfish to the preovulatory steroidal pheromone-17α20β-dihydroxy-4-pregnen-3-one. *J Comp Physiol (A)* **166**, 373–383.

Sosinsky A., Glusman G. and Lancet D. (2000). The genomic structure of human olfactory receptor genes. *Genomics* **70**, 49–61.

Soucek G., Breit S., Konig H. and Liebich H. (1999). Functional significance of musculature of the external nose in swine (*Sus scrofa f. domestica*). *Anat Histol Embryol* **28**, 307–314.

Speca D., Lin D.M., Sorensen P.W., Isacoff E.Y., *et al.* (1999). Functional identification of a goldfish odorant receptor. *Neuron* **23**, 487–498.

Spielman A., Sunavala G., Harmony J., Stuart W., *et al.* (1998). Identification and immunohistochemical localization of protein precursors to human axillary odors in apocrine glands and secretions. *Arch Dermatol* **134**, 813–818.

Spielman A., Zeng X., Leyden J. and Preti G. (1995). Proteinaceous precursors of human axillary odor — isolation of two novel odorant binding proteins. *Experientia* **51**, 40–47.

Stahlbaum C.C. and Houpt K.A. (1989). The role of the flehmen response in the behavioral repertoire of the stallion. *Physiol Behav* **45**, 1207–1214.

Starck D. (1960). Das Cranium eines Schimpansenfetus (*Pan trogdolytes*) von 71mm SchStgl. *Morph Jb* **38**, 559–647.

Starck D. (1984). The nasal cavity and nasal skeleton of Tarsius. In: *Biology of Tarsiers* (Niemitz C., ed.). G. Fischer, Stuttgart, pp. 275–290.

Steel E. and Keverne E.B. (1985). Effect of female odour on male hamsters mediated by the vomeronasal organ. *Physiol Behav* **35**, 195–200.

Stephan H., Baron, G. and Frahm, H. (1982). Comparison of brain structure volumes in Insectivora and Primates, II: Accessory olfactory bulb. *J Hirnforsch* **23**, 575–591.

Stephan H. and Kuhn H. (1982). The brain of *Micropotamagale lamotti*. *Z Saugetierk* **47**, 129–142.

Stevens K., Perry G.C. and Long S.E. (1982). Effect of ewe urine and vaginal secretions on ram investigative behaviour. *J Chem Ecol* **8**, 23–29.

Stodart E. (1966). Management and behaviour of breeding groups of the marsupial *Perameles nastua* in captivity. *Aust J Zool* **14**, 611–623.

Stone A. and Holtzman D.A. (1996). Feeding responses in young boa constrictors are mediated by the vomeronasal system. *Anim Behav* **52**, 949–955.

Stonerook M.J. and Harder J.D. (1992). Sexual maturation in female grey short-tailed opossums *M. domestica*, is dependent upon male stimuli. *Biol Reprod* **46**, 290–294.

Stralendorf F. von. (1986). Urinary chemosignals and specific behavioural responses in Tree Shrews. *J Chem Ecol* **12**, 99–106.

Strausfeld N. and Hildebrand J. (1999). Olfactory systems: common design, uncommon origins? *Curr Opin Neurobiol* **9**, 634–639.

Strotmann J., Hoppe R., Conzelmann S., Feinstein P., *et al.* (1999). Small subfamily of olfactory receptor genes: structural features, expression pattern and genomic organization. *Gene* **236**, 281–291.

Strotmann J., Wanner I., Helfrich T. and Breer H. (1995). Receptor expression in olfactory neurons during rat development: *in situ* hybridization studies. *Eur J Neurosci* **7**, 492–500.

Sugai T., Sugitani M. and Onoda N. (2000). Novel subdivisions of the rat accessory olfactory bulb revealed by the combined method with lectin histochemistry, electrophysiological and optical recordings. *Neuroscience* **95**, 23–32.

Sussman R. and Raven P. (1978). Pollination by lemurs and marsupials: an archaic coevolutionary system. *Science* **200**, 731–736.

Suzuki Y., Takeda M. and Obara N. (1998). Colchicine-induced cell death and proliferation in the olfactory epithelium and vomeronasal organ of the mouse. *Anat Embryol* **198**, 43–51.

Swann J. and Fiber J. (1997). Sex differences in function of a pheromonally stimulated pathway: role of steroids and the main olfactory system. *Brain Res Bull* **44**, 409–413.

Szabo K. and Mendoza A. (1988). Developmental studies on the rat vomeronasal organ: vascular pattern and neuroepithelial differentiation. I. Light microscopy. *Brain Res* **467**, 253–258.

Takahashi S., Iwanaga T., Takahashi Y., Nakano Y., *et al.* (1984). Neuron-specific enolase, neurofilament protein and S-100 protein protein in the olfactory mucosa of human fetuses: an immunohistochemical study. *Cell Tiss Res* **238**, 231–234.

Takami S., Getchell M. and Getchell T. (1995). Resolution of sensory and mucoid glycoconjugates with terminal α-galactose residues in the mucomicrovillar complex of the vomeronasal sensory epithelium by dual confocal laser-scanning microscopy. *Cell Tiss Res* **280**, 211–216.

Takami S., Getchell, M. and Chen, Y. (1993). Neurone-specific compounds in the receptor cells of the adult human vomeronasal organ. *Neuroreport* **4**, 375–378.

Takami S., Graziadei P. and Ichikawa M. (1992). The differential staining patterns of two lectins in the accessory olfactory bulb of the rat. *Brain Res* **598**, 337–342.

Takami S. and Hirosawa K. (1987). Light microscopic observations of the vomeronasal organ of Habu, *Trimeresurus flavoviridis*, Ophidia. *J Exp Med* **57**, 163–174.

Takami S., Luer C. and Graziadei P. (1994). Microscopic structure of the olfactory organ of the Clearnose Skate, *Raja eglanteria. Anat Embryol (Berl)* **190**, 211–230.

Takigami S., Mori Y. and Ichikawa M. (2000). Projection pattern of vomeronasal neurons to the accessory olfactory bulb in goats. *Chem Senses* **25**, 387–393.

Takigami S., Osada T., Matsuoka J., Matsuoka M., *et al.* (1999). The expressed localization of rat putative pheromone receptors. *Neuroscience Lett* **272**, 115–118.

Tanaka M., Treloar H., Kalb R.G., Greer C.A., *et al.* (1999). G(o) protein-dependent survival of primary accessory olfactory neurons. *Proc Natl Acad Sci* **96**, 14106–14111.

Taniguchi K., Arai T. and Ogawa K. (1993). Fine-structure of the septal olfactory Organ of Masera, and its associated gland in the Golden Hamster. *J Vet Med Sci* **5**, 107–116.

Taniguchi K., Matsusaki Y., Ogawa K. and Saito T.R. (1992). Fine structure of the vomeronasal organ in the Common Marmoset, *Callithrix jacchus. Folia Primatol (Basel)* **59**, 169–176.

Taniguchi K. and Mikami S. (1985). Fine structure of the epithelia of vomeronasal organ in horses and cattle: a comparative study. *Cell Tiss Res* **240**, 41–48.

Taniguchi K., Nii Y. and Ogawa K. (1993). Subdivisions of the accessory olfactory bulb as demonstrated by lectin histochemistry in the golden hamster. *Neurosci Lett* **158**, 185–188.

Taniguchi K., Toshima Y. and Saito T. (1996). Development of the olfactory epithelium and vomeronasal organ in the Japanese reddish frog. *Rana japonica. J Vet Med Sci* **58**, 7–15.

Taniguchi M., Wang D. and Halpern M. (1998). The characteristics of the electro-vomeronasogram: its loss following vomeronasal axotomy in the garter snake. *Chem Senses* **23**, 653–659.

Taniguchi M., Wang D. and Halpern M. (2000). Chemosensitive conductance and inositol 1,4,5-trisphosphate-induced conductance in snake vomeronasal receptor neurons. *Chem Senses* **25**, 67–76.

Tarozzo G., Peretto P., Perroteau I., Andreone C., *et al.* (1994). GnRH neurons and other cell populations migrating from the olfactory neuroepithelium. *Ann d'Endocrinol* **55**, 249–254.

Tarozzo G., Cappello P., DeAndrea M., Walters E., *et al.* (1998). Prenatal differentiation of mouse vomeronasal neurones. *Eur J Neurosci* **10**, 392–396.

Tavolga W. (1956). Visual, chemical and sound stimuli as cues in the sex discriminatory behavior of the gobiid fish *Bathygobius soporator. Zoologica* **41**, 49–65.

Tegoni M., Pelosi P., Vincent F., Spinelli S., *et al.* (2000). Mammalian odorant binding proteins. *Biochim Biophys Acta* **1482**, 229–240.

Teicher M. and Blass E. (1976). Suckling in new-born rats, eliminated by nipple-lavage, restored by saliva. *Science* **193**, 422–425.

Terman C.R. (1984). Sexual maturation of male and female White-footed Mice: influence of physical or urine contact with adults. *J Mammal* **65**, 97–102.

Terrick T.D., Mumme R.L. and Burghardt G.M. (1995). Aposematic coloration enchances chemosensory recognition of noxious prey in the garter snake *Thamnophis radix. Anim Behav* **49**, 857–866.

Thavathiru E., Jana N. and De P.K. (1999). Abundant secretory lipocalins displaying male and lactation-specific expression in adult hamster submandibular gland: cDNA cloning and sex hormone-regulated repression. *Eur J Biochem* **266**, 467–476.

Thiesen D., Regnier F., Rice M., Goodwin M., *et al.* (1974). Identification of a ventral scent marking pheromone in the male Mongolian Gerbil (*Meriones unguiculatus*). *Science* **184**, 83–85.

Thompson K.V. (1991). Flehmen and social-dominance in captive female Sable Antelope, *Hippotragus niger. Appl Anim Behav Sci* **29**, 121–133.

Thompson K.V. (1995a). Flehmen and birth synchrony among female Sable Antelope, *Hippotragus niger. Anim Behav* **50**, 475–484.

Thompson K.V. (1995b). Ontogeny of flehmen in Sable Antelope, *Hippotragus niger. Ethology* **101**, 213–221.

Thor D.H. (1978). Gonadal testosterone and perineal sniffing behavior of the male rat. *Behav Biol* **24**, 256–264.

Thornhill R.A. (1972). Ultrastructure of the accessory olfactory organ in the River Lamprey (*Lampetra fluviatilis*). *Acta Zool* **53**, 49–56.

Tirindelli R., Mucignat-Caretta C. and Ryba J. (1998). Molecular aspects of pheromonal communication via the vomeronasal organ of mammals. *Trends Neurosci* **21**, 482–486.

Tobet S.A., Crandall J. and Schwarting G. (1992). Relationship of migrating LHRH neurons to unique olfactory system glycoconjugates in embryonic rats. *Dev Biol* **155**, 471–482.

Tobet S.A., Sower S. and Schwarting G. (1997). Gonadotropin-releasing hormone containing neurons and olfactory fibers during development: from lamprey to mammals. *Brain Res Bull* **44**, 479–486.

Toftegaard C., Moore C. and Bradley A.J. (1999). Chemical characterisation of urinary pheromones in Brown Antechinus, *Antechinus stuartii*. *J Chem Ecol* **25**, 527–535.

Toubeau G., Cotman C. and Bels V. (1994). Morphological and kinematic study of the tongue and buccal cavity in the lizard *Anguis fragilis*. *Anat Rec* **240**, 423–433.

Toyoda F. and Kikuyama S. (2000). Hormonal influence on the olfactory response to a female-attracting pheromone, sodefrin, in the newt, *Cynops pyrrhogaste*. *Comp Biochem Physiol [B]* **126**, 239–245.

Tristram D.A. (1977). Intraspecific olfactory communication in the terrestrial salamander *Plethoden cinereus*. *Copeia* 597–600.

Troemel E. (1999). Chemosensory signalling in *C. elegans*. *BioEssays* **21**, 1011–1020.

Trotier D. and Døving K. (1996). Inward rectifiying current is activated by hyperpolarisation in Frog vomeronasal receptor cells. *Prim Sensory Neuron* **1**, 245–261.

Trotier D. and Doving K. (1998). "Anatomical description of a new organ in the nose of domesticated animals" by Ludvig Jacobson (1813). *Chem Senses* **23**, 743–754.

Trotier D., Døving K., Ore K. and Shalchian-Tabrizi C. (1998). Scanning electron microscopy and gramicidin patch clamp recordings of microvillous receptor neurons dissociated from the rat vomeronasal organ. *Chem Senses* **23**, 49–57.

Trotier D., Døving K.B. and Rosin J.-F. (1994). Functional properties of frog vomeronasal receptor cells. In: *Olfaction and Taste* **XI** (Kurihara K., *et al.*, eds.). Springer, Berlin, pp. 188–191.

Trotier D., Eloit C., Wassef M., Talmain G., *et al.* (2000). The vomeronasal cavity in adult humans. *Chem Senses* **25**, 369–380.

Tsujikawa K. and Kashiwayanagi M. (1999). Protease-sensitive urinary pheromones induce region-specific fos-expression in rat accessory olfactory bulb. *Biochem Biophys Res Comm* **260**, 222–224.

Tubbiola M. and Wysocki C.J. (1997). FOS-ir following exposure to conspecific or heterospecific urine: where are chemosensory cues sorted? *Physiol Behav* **62**, 867–870.

Tucker D. (1971). Non-olfactory responses from nasal cavity: Jacobson's Organ and trigeminal system. In: *Handbook of Sensory Physiology: Chemical Senses, 1. Olfaction* (Biedler L., ed.). Springer, Berlin, pp. 151–181.

Turin L. (1996). A spectroscopic mechanism for primary olfactory reception. *Chem Senses* **21**, 773–791.

Ulibarri C. and Yahr P. (1996). Effects of androgens and estrogens on sexual differentiation of sex behavior, scent marking and the sexually dimorphic area of the gerbil hypothalamus. *Horm Behav* **30**, 107–130.

Utsumi M., Ohno K., Kawasaki Y., Tamura M., *et al.* (1999). Expression of major urinary protein genes in the nasal glands associated with general olfaction. *J Neurobiol* **39**, 227–236.

Valencia A., Collado P., Cales J.M., *et al.* (1992). Postnatal administration of dihydrotestosterone to the male rat abolishes sexual dimorphism in the accessory olfactory bulb: a volumetric study. *Dev Brain Res* **68**, 132–135.

van der Lee S. and Boot L. (1955). Spontaneous pseudo-pregnancy in mice. *Acta Physiol Pharm (Neerl)* **4**, 442–444.

van Groen T., Ruardy L. and Dasilva F. (1986). Electrophysiological mapping of the accessory olfactory bulb of the rabbit (*Oryctolagus cuniculus*). *Exp Neurol* **93**, 67–76.

Vandenbergh J.G. (1987). Regulation of puberty and its consequences on population dynamics in mice. *Amer Zool* **27**, 891–898.

Vandenbergh J.G. (1989). Coordination of chemical signals and ovarian function during sexual development. *J Anim Sci* **67**, 1841–1847.

Vandenbergh J., Whitsett J. and Lombardi J. (1975). Partial isolation of a pheromone accelerating puberty in female mice. *J Reprod Fertil* **43**, 515–523.

Vasilieva N., Lai S. and Wysocki C.J. (1995). Role of the vomeronasal organ in sociosexual behavior in male Djungarian hamsters. *Chem Senses* **20**, 128.

Verberne G. (1976). Chemo-communication among domestic cats mediated by the olfactory and vomero nasal senses, 2: the relation between the function of Jacobson's [vomeronasal] organ and flehmen behaviour. *Z Tierpsychol* **42**, 113–128.

Verheijen F. and Reutte R. (1969). The effect of alarm substances on predation among cyprodinids. *Anim Behav* **17**, 551–554.

Vincent F., Spinelli S., Ramoni R., Grolli S., *et al.* (2000). Complexes of porcine odorant binding protein with odorant molecules belonging to different chemical classes. *J Mol Biol* **300**, 127–139.

Vincent S. and Kimura H. (1992). Histochemical mapping of NO-synthase in rat brain. *Neuroscience* **46**, 755–784.

von Brunn A. (1892). Beiträge zur mikroskopischen Anatomie des menschlichen Nasenhöhle. *Arch Mikro Anat* **39**, 632.

von Campenhausen H. and Mori K. (2000). Convergence of segregated pheromonal pathways from the accessory olfactory bulb to the cortex in the mouse. *Eur J Neurosci* **12**, 33–46.

von Campenhausen H., Yoshihara Y. and Mori K. (1997). OCAM reveals segregated mitral/tufted cell pathways in developing accessory olfactory bulb. *Neuroreport* **8**, 2607–2612.

von Eerdenburg F. and Swaab D. (1994). Postnatal development and sexual differentiation of pig hypothalamic nuclei. *Psychoneuroendcrinol* **19**, 471–484.

von Helversen O., Winkle R.L. and Bestmann H.J. (2000). Sulphur-containing "perfumes" attract flower-visiting bats. *J Comp Physiol [A]* **186**, 143–153.

von Mihalkovics V. (1898). Nasenhöle und Jacobsonsches Organ. Eine morphologische Studie. *Anat Hefte* **1**(xi), 1–107.

von Navratil D. (1926). Uber das Jacobsonche Organ der Wirbeltiere. *Z Anat Entw Gesch*, 648–656.

von Rekowski C. and Zippel H. (1997). Delayed functional recovery for pheromone recognition after bilateral olfactory nerve section in Goldfish. *Chem Senses* **22**, 235 (abs. 143).

von Uexkull J. and Sarris E. (1931). Das Duftfeldt des Hundes. *Z Hundefors* **1**, 55–68.

Wabnitz P., Bowie J., Tyler M., Wallace J. and Smith B. (1999). Aquatic sex pheromone from a male tree frog. *Nature* **401**, 444–445.

Wallace P. (1977). Individual discrimination of humans by odor. *Physiol Behav* **19**, 577–579.

Waldman B. (1985). Olfactory basis of kin-recognition in Toad tadpoles. *J Comp Physiol A* **156**, 565–567.

Wang F., Nemes A., Mendelsohn M. and Axel R. (1998). Odorant receptors govern the formation of a precise topographic map. *Cell* **93**, 47–60.

Wang R.T. and Halpern M. (1980a). Light and electron microscopic observations on normal structure of vomeronasal of Garter snakes. *Morphol Jb* **164**, 47–68.

Wang R.T. and Halpern M. (1980b). Scanning electron microscopic studies of the surface morphology of the vomeronasal epithelium and olfactory epithelium of Garter snakes. *J Anat* **157**, 399–428.

Wang D., Jiang X., Chen P., Inouchi J., *et al.* (1993). Chemical and immunological analysis of prey-derived vomeronasal stimulants. *Brain Behav Evol* **41**, 246–254.

Washabaugh K. and Snowdon C.T. (1998). Chemical communication of reproductive status in female Cotton-Top Tamarins (*Saguinus oedipus oedipus*). *Am J Primatol* **45**, 337–349.

Wattiez R., Remy C., Falmagne P. and Toubeau G. (1994). Characterization of a frog-derived proteinaceous chemoattractant eliciting prey attack by Checkered Garter Snakes (*Thamnophis marcianus*). *J Chem Ecol* **20**, 1143–1146.

Wedekind C. and Furi S. (1997). Body odour preferences in men and women: do they aim for specific MHC combinations or simply heterozygosity? *Proc Royal Soc Lond B* **264**, 1471–1479.

Weiler E., Apfelbach R. and Farbman A.I. (1999). The vomeronasal organ of the male ferret. *Chem Senses* **24**, 127–136.

Weiler E., McCulloch M.A. and Farbman A.I. (1999). Proliferation in the vomeronasal organ of the rat during postnatal development. *Europ J Neurosci* **11**, 700–711.

Wekesa K. and Anholt R. (1997). Pheromone regulated production of inositol-(1,4,5)-trisphosphate in the mammalian vomeronasal organ. *Endocrinology* **138**, 3497–3504.

Wekesa K. and Anholt R. (1999). Differential expression of G-proteins in the mouse olfactory system. *Brain Res* **837**, 117–126.

Wekesa K. and Lepri J.J. (1994). Removal of the vomeronasal organ reduces reproductive performance and aggression in male Prairie voles. *Chem Senses* **19**, 35–45.

Weller L. and Weller A. (1993). Human menstrual synchrony: a critical assessment. *Neurosci BioBehav Rev* **17**, 427–439.

Welsh C.J., Moore R.E., Bartlett R.J. and Jackson, L.L. (1988). Novel species-typical esters from preputial glands of sympatric voles. *J Chem Ecol* **14**, 143–158.

Wemmer C. (1977). Comparative ethology of Lesser Spotted Genet, and some related viverrids. *Smiths Contrib Zool* **239**, 1–93.

Wesinger S.R. and Baum M. (1997). Sexually dimorphic processing of somato- and chemosensory inputs to forebrain LHRH neurons in mated ferrets. *Endocrinology* **138**, 1121–1129.

Westhofen M.C. and Herberhold C. (1987). Trigeminal and olfactory synergism in the perception of smell. *Acta Oto-Rhino-Laryngol Belg* **41**, 66–71.

Whitney G., Alpern M. and Dizinno G. (1974). Female odors evoke ultrasounds from male mice. *Anim Learn Behav* **2**, 13–18.

Wilcox R. and Johnson R.E. (1995). Scent counter-marks: specialised mechanisms of perception and response to odors in hamsters. *J Comp Psychol* **109**, 349–356.

Wildt L., Marshall G. and Knobil E. (1980). Experimental induction of puberty in the infantile rhesus monkey. *Science* **207**, 1373–1375.

Wilson D.B. and Hendrickx A.G. (1977). Quantitative aspects of proliferation in the nasal epithelium of the rhesus monkey embryo. *J Emb Exp Morphol* **38**, 217–226.

Wilson H. (1987). Female axillary secretions influence womens' menstrual cycles: a critique. *Horm Behav* **21**, 536–546.

Wilson K.C. and Raisman G. (1980). Age-related changes in the neurosensory epithelium of the mouse vomeronasal organ: extended period of post-natal growth in size and evidence for rapid cell turnover in the adult. *Brain Res* **185**, 103–113.

Wilson K.C. and Raisman G. (1981). Estimation of numbers of vomeronasal synapses in the glomerular layer of the accessory olfactory bulb of the mouse at different ages. *Brain Res* **205**, 245–254.

Winans S.S. and Powers J.B. (1977). Olfactory and vomeronasal deafferentation of male hamsters: histological and behavioral analyses. *Brain Res* **126**, 325–344.

Wirsig C.R. and Leonard C.M. (1986). ACh-ase and LHRH distinguish sub-populations of terminal nerve neurons. *Neuroscience* **19**, 719–740.

Wirsig C.R. and Leonard C.M. (1987). Terminal nerve damage impairs the mating behavior of the male hamster. *Brain Res* **417**, 293–303.

Wirsig-Wiechmann C.R. (1993b). *Nevus terminalis* lesions, 1. No effect on pheromonally induced testosterone surges in male hamster. *Physiol Behav* **53**, 252–255.

Wirsig-Wiechmann C.R. and Lepri J. (1991). LHRH-immunoreactive neurons in the pterygopalatine ganglia of voles. *Brain Res* **568**, 289–293.

Witkin J.W. (1985). LHRH in olfactory bulbs of primates. *Am J Primatol* **8**, 309–316.

Witkin J.W. (1987). Immunocytochemical demonstration of luteinizing-hormone-releasing hormone in optic nerve and nasal region of fetal rhesus macaque. *Neurosci Lett* **79**, 73–77.

Wöhrmann-Repenning A. (1977). Comparative study on nasal fossae of *Tupaia* and four Insectivores. *Anat Anz* **124**, 375–384.

Wöhrmann-Repenning A. (1978). Geschmacksknospen an der *papilla palatina* von *Tupaia glis* (Diard 1820), ihr Vorkommen und ihre Beziehungen zum Jacobsonschen Organ. *Morphol Jb* **124**, 375–384.

Wöhrmann-Repenning A. (1981a). Zur embryonalen und frühen postnatalen Entwicklung des Jacobsenschen Organs in Bezieheung *zum ductus nasopalatinus* bei der Ratte. *Zool Anz* **206**, 203–214.

Wöhrmann-Repenning A. (1981b). Die topographie der mundungen der Jacobsonsches Organe des kaninchens unter funktionalle aspekt. *Z Saugetierk* **46**, 273–279.

Wöhrmann-Repenning A. (1987). Zur Anatomie des Vomeronasal komplexes von *Elephantulus rozeti* (Macroscelidae). *Zool Anz* **218**, 1–8.

Wöhrmann-Repenning A. and Barth-Müller U. (1994). Functional anatomy of the vomeronasal complex in the embryonic development of the pig (*Sus scrofa*). *Acta Theriol* **39**, 313–323.

Won J., Mair E., Bolger W. and Conran R. (2000). The vomeronasal organ: an objective anatomic analysis of its prevalence. *ENT J* **79**, 600–605.

Wong M., Chen Y. and Moss R. (1993). Excitatory and inhibitory synaptic processing in the accessory olfactory system of the female rat. *Neurosci* **56**, 355–365.

Wood R.I. (1998). Integration of chemosensory and hormonal input in the male Syrian hamster brain. *Proc Natl Acad Sci* **855**, 362–372.

Wong S.T., Trinh K., Hacker B., Chan G.C.K. *et al.* (2000). Disruption of the Type III adenylyl cyclase gene leads to peripheral and behavioral anosmia in transgenic mice. *Neuron* **27**, 487–497.

Wray S., Grant P. and Gainer H. (1989). Evidence that cells expressing LHRH-mRNA in the mouse are derived from progenitor cells in the olfactory placode. *Proc Natl Acad Sci* **86**, 8132–8136.

Wright R. (1883). On the Organ of Jacobson in Ophidia. *Zool Anz* **6**, 389–393.

Wu Y., Tirindelli R, and Ryba N.J. (1996). Evidence for different chemo-sensory transduction pathways signal in olfactory and vomeronasal neurons. *Biochem Biophys Res Comm* **220**, 900–904.

Wysocki C.J. (1989). Vomeronasal chemoreception: its role in reproductive fitness and physiology. *Prog Neurol Neurobiol* **50**, 545–566.

Wysocki C.J., Bean N.J. and Beauchamp G. (1986). The mammalian vomeronasal system: its role in learning and social behaviors. In: *Chemical Signals in Vertebrates* **4** (Duvall D., *et al.*, eds.). Plenum, New York, pp. 471–485.

Wysocki C.J., Beauchamp G.K., Reidinger R.F. and Wellington J.L. (1985). Access of large and non-volatile molecules to the vomeronasal organ of mammals during social and feeding behaviors. *J Chem Ecol* **11**, 1147–1159.

Wysocki C.J., Katz Y. and Bernard R. (1983). Male vomeronasal organ mediates female induced testosterone surges in mice. *Biol Reprod* **28**, 917–922.

Wysocki C.J., Kruczek M., Wysocki L.M. and Lepri J. (1991). Activation of reproduction in nulliparous and primiparous voles is blocked by vomeronasal organ removal. *Biol Reprod* **45**, 611–616.

Wysocki C.J. and Lepri J. (1991). Consequences of removing the vomeronasal organ. *J Ster Biochem Molec Biol* **39**(4B), 661–669.

Wysocki C.J. and Meredith M. (1987). The vomeronasal system. In: *The Neurobiology of Taste and Smell* (Finger T. and Silver W., eds.). Wiley, New York, pp. 125–150.

Wysocki C.J. and Preti G. (1998). Pheromonal influences. *Arch Sex Behav* **27**, 627–634.

Wysocki C.J. and Wysocki L. (1995). Surgical removal of the vomeronasal organ and its verification. In: *Experimental Cell Biology of Taste and Olfaction* (Spielman A. and Brand J., eds.). CRC Press, Boca Raton, pp. 49–58.

Wysocki C.J., Zeng C. and Preti G. (1999). Specific anosmia and olfactory sensitivity to 3-methyl-2-hexanoic acid: a major component of human axillary odor. *Chem Senses* **25**, 652 (abs. 330).

Yamamoto K., Kawai Y., Hayashi T., Ohe Y., *et al.* (2000). Silefrin, a sodefrin-like pheromone in the abdominal gland of the sword-tailed newt, *Cynops ensicauda*. *FEBS Lett* **472**, 267–270.

Yamazaki K., Beauchamp G., Singer A., Bard J. and Boyse E. (1999). Odortypes: their origin and composition. *Proc Natl Acad Sci* **96**, 1522–1525.

Yokoi M., Mori K. and Nakanishi S. (1995). Refinement of odor molecule tuning by dendrodendritic synaptic inhibition in the olfactory bulb. *Proc Natl Acad Sci* **92**, 3371–3375.

Yokosuka M., Matsuoka M., Ohtani-Kaneko R., Iigo M., *et al.* (1999). Female-soiled bedding induced Fos immunoreactivity in the ventral part of the premammillary nucleus (PMv) of the male mouse. *Physiol Behav* **68**, 257–261.

Yoshida K., Tobet S., Crandall J., Jiminez T. and Schwarting G. (1995). The migration of LH-RH neurons in the developing rat is associated with a transient caudal projection of the vomeronasal nerve. *J Neurosci* **15**, 7769–7777.

Yoshida K., Rutishauser U., Crandall J. and Schwarting G. (1999). Polysialic acid facilitates migration of luteinizing hormone-releasing hormone neurons on vomeronasal axons. *J Neurosci* **19**, 794–801.

Yoshihara Y., Kawasaki M., Tamada A., Fujita H., *et al.* (1997). OCAM: a new member of the neural cell adhesion molecule family related to zone-to-zone projection of olfactory and vomeronasal axons. *J Neurosci* **17**, 5830–5842.

Yoshihara Y. and Mori K. (1997). Basic principles and molecular mechanisms of olfactory axon pathfinding. *Cell Tissue Res* **290**, 457–463.

Young B. (1993). Evaluating hypotheses for the transfer of stimulus particles to Jacobson's Organ in snakes. *Brain Behav Evol* **41**, 203–209.

Yukimatsu M., Takami S., Matsumura G. and Nishiyama F. (2000). Immunoreactivity for G-proteins in the vomeronasal organ of human fetuses. *Chem Senses* **25**, 215 (abs. P-17).

Zancanaro C., Caretta C., Bolner A., Sbarbati A., *et al.* (1997). Biogenic amines in the vomeronasal organ. *Chem Senses* **22**, 439–445.

Zancanaro C., Mucignat-Caretta C., Merigo F. and Cavaggioni A. (1999). Immunohistochemical investigation of the vomeronasal organ. Nitric oxide synthase expression in the mouse during postnatal development. *Neurosci Lett* **269**, 5–8.

Zancanaro C., Mucignat-Caretta C., Merigo F. and Osculati F. (1999). Neuropeptide expression in the mouse vomeronasal organ during postnatal development. *Neuroreport* **10**, 2023–2027.

Zbar R., Zbar L., Dudley C., Trott S., *et al.* (2000). A classification schema for the vomeronasal organ in humans. *Plast Reconstr Surg* **105**, 1284–1288.

Zeiske E.B., Melinkat R., Breucker H. and Kux J. (1976). Ultrastructural studies on the epithelia of the olfactory organ of cyprinodonts (Teleostei, Cyprinodontoidea). *Cell Tissue Res* **172**, 245–267.

Zeiske E.B., Theisen B. and Breuckner H. (1989). Olfactory organs in pelagic and benthic elasmobranchs. *Zool Fortsch* **35**, 370–372.

Zeng C., Leyden J., Spielman A. and Preti G. (1996). Analysis of characteristic human female axillary odors — qualitative comparison to males. *J Chem Ecol* **22**, 237–257.

Zeng C., Vowels B., Spielman A., Leyden J., *et al.* (1996). A human axillary odorant is carried by Apolipoprotein-D. *Proc Natl Acad Sci* **93**, 6626–6630.

Zidek L., Stone M., Lato S., Pagel M., *et al.* (1999). NMR-mapping of the recombinant mouse major urinary protein-I binding site occupied by the pheromone 2-sec-butyl-4,5-dihydrothiazole. *Biochem* **38**, 9850–9861.

Zingeser M.R. (1984). The nasopalatine ducts and associated structures in the rhesus monkey (*Macaca mulatta*) — topography, prenatal development, function and phylogeny. *Am J Anat* **170**, 581–595.

Zippel H., Gloger M., Luthje L., Nasser S., *et al.* (2000). Pheromone discrimination ability of olfactory bulb mitral and ruffed cells in the goldfish. *Chem Senses* **25**, 339–349.

Zuri I., Fishelson L. and Terkel J. (1998). Morphology and cytology of the nasal cavity and vomeronasal organ in juvenile and adult blind Mole rats (*Spalax ehrenbergi*). *Anat Rec* **251**, 460–471.

INDEX